U.S. ARMY SIGNAL CORPS VEHICLES 1939–45

CASEMATE | ILLUSTRATED | SPECIAL

C CASEMATE | ILLUSTRATED | SPECIAL

U.S. ARMY SIGNAL CORPS VEHICLES 1939–45

DIDIER ANDRES

CISS0008

Published in the United States of America and Great Britain in 2021 by
CASEMATE PUBLISHERS
1950 Lawrence Road, Havertown, PA 19083, USA
and
The Old Music Hall, 106–108 Cowley Road, Oxford OX4 1JE, UK

This book is published in cooperation with, and under license from, Sophie Histoire
& Collections. Originally published in French as Les Véhicules des transmissions
de l'U.S. army © Histoire & Collections 2020

Hardback Edition: ISBN 978-1-63624-064-0
Digital Edition: ISBN 978-1-63624-065-7

A CIP record for this book is available from the British Library

Design by Myriam Bell
Translated by Alan McKay
Printed and Bound in Turkey by Megaprint

For a complete list of Casemate titles, please contact:

CASEMATE PUBLISHERS (US)
Telephone (610) 853-9131
Fax (610) 853-9146
Email: casemate@casematepublishers.com
www.casematepublishers.com

CASEMATE PUBLISHERS (UK)
Telephone (01865) 241249
Email: casemate-uk@casematepublishers.co.uk
www.casematepublishers.co.uk

Page 2 image: At Haleiwa, Hawaii, an SCR-547 optical heightfinder radar in
position for surveillance, partnering an SCR-268 mobile long-wave searchlight
control set visible behind it. An M1 telemeter added to the precision of the
work. (U.S. Army Air Forces Ref. 65278 A.C.)

Contents page image: Germany, 1945, the 40th Mobile Command Unit
forms up its column to continue its advance into enemy territory. (Private
collection)

Contents

1

Introduction

As a vital part of the U.S. Army, the Signal Corps was at the forefront of the technological development of communications throughout World War II. Created well before the 1940s, its roots went back to just before 1870 when Congress made its existence official.

Reducing the job of the Signal Corps to simply laying telephone cables is confining it to a far too limited role. On December 26, 1944, at Monghidoro, Italy, Sergeant S. Gerbert and Private William Williams from the 53rd Signal Battalion are testing a vital hookup. (U.S. Army SC-199368)

Although operational during World War I, it was in 1937 that the Signal Corps really got off the ground, at the same time as the early developments of radar. After he was appointed head of the Signal Corps, Major General Spencer B. Akin, Chief Signal Officer from 1947 to 1951, praised the service and its amazing flexibility:

"It is indeed paradoxical that anything good could come out of the hideous and awful tragedy of the last war.

"Yet it is so. It is, of course, true that any such paradoxical byproduct cannot even to begin to balance the scales so heavily weighted on the one end with the want, suffering, and human misery that war brought.

Above: All throughout the conflict millions of photos were taken, often in terrible conditions. The Signal Corps companies were active on all the fronts, like here on D-Day+3, where Private First Class Josiah S. Carpenter has just arrived in Normandy with the 165th Photographic Signal Company. (U.S. Army SC-380290)

Nevertheless, the conflict hastened progress in many fields, and its exigencies quickly brought developments that might otherwise have been many years in arriving.

"The Army Signal Corp was in one of those fields. Out of this Corps' experience in the war years came a scientific progress that has left its indelible mark on the world.

"More than that, the benefits of these developments are enjoyed daily by the people of the nation, for much of the progress has been routed from wartime to peacetime use.

"Whether it be radar probing beyond the limits of man's vision, frequency modulation (FM) radio sets in the living rooms of the country, electronics equipment plotting the course and direction of winds 60,000 feet up and forecasting the weather—whether it be these or scores of other scientific developments that have made life better and living easier, they are some of the results of Signal Corp activity.

"Of course, the Signal Corp did not accomplish all these things alone. Progress came as the result of teamwork on the part of science, industry, and the military. Some of this teamwork occurred before the

Top right: Thousands of documentary training films were made. The Pictorial Service of the Signal Corps was given this task. At the Holabird Quartermaster Depot, a film is being made on how to check and repair the electrical system of an engine. (U.S. Army SC-126382)

Right: Fort Monmouth, June 1941, the headquarters of the Signal Corps. New camera operators take charge of equipment which has just been issued to them. At this time in the early 1940s, vehicles were omnipresent, and the commencement of military operations emphasized the fact that troops had now become mechanized. (U.S. Army SC-120531)

war—and therein lies a lesson that must always be remembered if this nation is to retain its supremacy in communications: progress can be made only after painstaking research, sometimes after many tears of probing in a given field. Science, supported in peacetime, pays its dividends when war demands quick results.

"Thus, it becomes quite clear that many of the communications 'miracles' of World War II were not miracles at all. Though the war speeded up development and progress, a great many of these so-called miracles were the products of many years of earlier study.

"Today, remembering this lesson the war taught us, the Signal Corp is pressing a research and development program aimed at supplying integrated communications that are global in scope and capable of coordinating a far-flung defense in an age of vastly increasing speeds. This program, still characterized by teamwork, seeks to assure progress from the point where the tremendous achievements of wartime ended. Precedent assures it success. It was the men of the Signal Corp who stretched an 800,000-mile wartime radio and wire net over the globe—a network with a daily capacity of a hundred million words. Most of them had the spirit of a Signal Corp man General Patton once found on top of a telephone pole in Africa. The Allied drive toward Tunis was in full swing, and the soldier, making a fair target, was in the midst of furious fire. The General stopped and asked what he was doing up these at such a time. In colorful language which the colorful General could understand and appreciate, the soldier let him know that fire or no fire, communications wire had to be fixed. General Patton went away convinced that this was the "bravest man" he saw in the African campaign. Signals Corp tradition is and always has been to get the message through.

Top: Spring maneuvers at Fort Benning on April 23, 1942. The Signal Corps went to war with completely new equipment. Although communication-linked technological means were in full swing, transport was as important. This 161st Signal Photographic Company column is made up of Dodge Pick-up and Carryall 4×4, ½-ton trucks, Chevrolet Cargo and Panel 4×4, 1½-ton trucks, and GMC CCKW 6×6, 2½-ton trucks, all factory-new. (U.S. Army SC-135280)

Middle: Manila, Philippines, August 1, 1945, in the middle of a Signal Corps telephone cable depot. As well as doing its own work, the Corps also had to look after the supplies and consumables needed by the troops. (U.S. Army SC-312962)

Left: These depots were everywhere in the United States and in all the conflict zones. Like other army provisions, Signal Corps supplies were meticulously cared for and stored in a particularly orderly manner. (U.S. Army SC-312961)

"As in all branches, there was a tremendous personnel growth in the Signal Corp during the war years. In September of 1940, the Corps has 282 officers and 6,261 enlisted men. In March of 1945, it had 27,848 officers and 503,257 enlisted men. They built, operated, and maintained all kinds of communications. They supplied the entire Army with more than 100,000 signal items. They made the Army's pictures and photographed its V-mail. They forecast the weather. They pioneered in electronics development. Overlooking no means of transmitting a message, they trained and cared for pigeons. And constantly, the specialists of the Signal Corp trained new specialists and prepared the technical manuals that led to an understanding of equipment.

"Communications is the primary mission of the Signal Corp. The faster the pace of war and the greater the distances involved, the more critically important are quick, reliable, flexible, and secure means of transmitting messages. The Signal Corp had to provide the means of co-ordinating every American military activity in World War II.

"To co-ordinate a global war required first of all a global communications network. One was built, with 800,000 miles of circuits. A 10-word printed message could be sent around the world in nine and a half seconds. The same net made it possible for President Roosevelt to have a White House map changed every hour, showing the war situation down to battalion level. It made it possible for him to beat Prime Minister Winston Churchill by two minutes in announcing the Italian surrender. While at Yalta, the President—through Signal Corp facilities hastily installed—had many teletype conferences with General Eisenhower in France, with the White House and the State Department in Washington, and with Ambassador Hurley in China. That globe-girdling net included some historic construction, like the 2,060-mile line along the Alaska Military Highway, and the Army Airways Communications System that furnished 'radio highways' for the Air Transport Command.

"The war brought some significant developments in communications. Radio relay equipment was designed

Top right: The Jeep was appreciated by all its users, especially the Signal Corps. As it was not specifically designated as a Corps vehicle, it had no "K" designation. (U.S. Army SC-278)

Right: Whether it was the ¾-ton or ½-ton (as here), the Dodge Carryall was one of the Signal Corps' favorite vehicles. It too did not qualify for the "K" designation. At Fort Benning, Sergeant William F. Cox of the 161st Signal Photographic Company supervises Private Max Nahman and Private First Class Edward Jankowski loading a vehicle roof rack, February 19, 1942. (U.S. Army SC-132147)

to span breaks in wire transmission caused by rapid movement of troops. It proved so dependable that, in many cases, wire was not installed, even in stabilized situations.

"It was the wartime efforts of the Signal Corp that gave impetus to the development of frequency modulated (FM) radio. Switchboards were simplified and lightened. A light-weight field telephone that required no batteries was built. Power for the speaking circuit was generated by the speaker's voice. The war brought the walkie-talkie and the handie-talkie. It also brought many new types of wire and cable. Assault wire became so light that one man could carry a mile of it. New dispensers were developed, so that coiled wire could be paid out from the packboard of a soldier on foot, from the rear of a jeep, from a light airplane flying at two miles a minute over impenetrable jungle, or even from a rocket or grenade in flight.

"From the procurement standpoint, wartime Signal Corp activities constituted a six and a half billion-dollar business. While radio and wire circuits pushed across the world's battlefields, at home the electronics industry grew from a prewar annual capacity of a quarter-billion dollars to one of three billion. More than 200 plant expansions were financed in 20 states. Production of quartz crystals, important in stabilizing radio frequencies, increased 10 times. Dry battery output in the first quarter of 1943 was 100,000,000 cells; in the first quarter of 1945, it was 623,000,000 cells. Every month our Army used 250,000 miles of assault and field wire. Because we were indeed fighting a world war, countless pieces of delicate equipment had to be treated for the cold of Alaska and the damp heat of the South Pacific.

"Photographically, the Signal Corp made both still and motion pictures for tactical, strategic, intelligence, technical, and training activities in prosecuting the war; and for news and historical uses (such as most of the pictures illustrating this volume) by the public. Hundreds of thousands of still pictures were made. There were 2,659 training films produced during the war, and 302,000 duplicate prints of them were distributed through Signal Corp libraries. The Signal Corp also distributed to overseas based the entertainments films that Hollywood donated during the war. Projection equipment was procured, maintained, and loaned by the Signal Corp, which also trained the projectionists. As a related activity, thousands of film strips were produced

On parade for an inspection, this homing pigeon unit displays its vehicles. The fleet consists of two Slat-Grill Willys Jeeps, two Dodges, five K-70 Chevrolet panel trucks, two pigeon loft trailers, and seven Chevrolet cargo trucks. (U.S. Army Signal Corps)

and distributed to assist in training. In a purely war-born innovation, our photographic personnel processed one a quarter-billion V-mail letters between June, 1942 and August, 1945.

"Accurate weather predictions played an important part in achieving victory. In the course of its meteorological work, the Signal Corp developed radio equipment by which the direction and velocity of the wind could be plotted at elevations up to 60,000 feet.

"In another realm of research, the Signal Corp developed nearly all important radar equipment used in World War II combat, before work in that field became an Air Force responsibility. The Corps also aided in developing proximity fuses.

"When wire communication was not available and radio silence had to be maintained, the Signal Corps turned to carrier pigeons. Oldest among the birds was Kaiser, who was captured from the Germans in World War I, who transferred his allegiance to the United States, and served again in World War II. There was also Jungle Joe, who carried valuable information

over some of the steepest mountains in Asia; Burma Queen, who was dropped from a B-25 and then flew 320 miles over Burma mountains; and Yank, who in Africa flew 90 miles in 100 minutes with a message from General Patton.

"Nothing is too old or too new for the Signal Corps if—under a given set of conditions—it will do the best job of delivering a message. For the Signal Corp is the Army's voice of command, a voice that must always speak if the nation is to defend itself successfully."

It was in this spirit that Major General S. B. Akin expressed himself to describe, in broad outline, a service in the armed forces. But what doesn't come through from reading this report is the extreme precision of the Signal Corps' demands. As for the vehicles—which interest us here—almost every large piece of equipment at the start of the conflict was going to need wheeled transport for mobility. With time, standardization appeared, with "K"-designated materiel providing a specific idea of the Signal Corps' mobile means.

Brooklyn Army Base, August 22, 1942. Amid the civilian vehicles and U.S. Army Air Forces tankers, the equipment for an SCR-268 radar and an SCR-270 radar is ready to be loaded for overseas. Note the K-28 and K-22 trailers as well as the K-39, K-31, K-32, and K-33 trucks. (U.S. Army SC-126014)

2

Evolution of the Model K

The Model "K" designation was unique and was specific to the Signal Corps, in the same way that all the radio equipment was defined by the prefix SCR (Signal Corps Radio) to which was added a three-figure number.

The date when the "Model K" category of materiel was created is not mentioned anywhere. It can, however, be placed more or less at the end of the 19th and the beginning of the 20th centuries. The idea was for the prefix "K" to designate a type of vehicle with a particular function, not its shape or manufacturer. When road vehicles appeared with their particular type of axles, only the primary task and aim was designated by a model in the "K" category.

When it was created, the Signal Corps' means of transmission, like the telephone, was limited, as was its transport, which consisted of one- or two-axled horse-drawn wagons. It wasn't until the mid-1930s that the internal combustion engine replaced the horse. Until 1939/40, the Model K category only comprised 17 vehicles, from the wagon to the Model-T Ford. The way wireless took off, developing over such a short time, meant the number of specialized vehicles and trailers increased, with the Model K list rising to 84 in quick time.

Administratively, the "K" designations were abandoned after February 16, 1943 and any new materiel was thereafter allocated a "V" number. This did not change things fundamentally for the rest of the

This picture taken at the Field Museum of Chicago in 1944 symbolizes the technological evolution of the Signal Corps: the Civil War with signals by messenger, World War I with its DIY cars, and World War II with its specialist vehicles like the K-51 Chevrolet. (Private collection)

In 1904, the telegraph had an important place in the communications system, with vehicles specially designed to connect it to the existing landline networks being designed. These were rarely designated as "K" models. (Private collection)

war as only a few vehicles and trailers were given the new appellation. On the other hand, vehicles the Signal Corps cataloged as Model Ks lost this designation very quickly. Vehicles for towing semi-trailers became common to all the armed forces and were no longer considered as specific to the Signal Corps.

Not counting the period preceding World War II, and those "K" numbers not attributed on the general

By the end of the 1930s the Signal Corps was using a wide array of vehicles, not necessarily K models which was a practice that disappeared with the onset of World War II. Here a 1937 van-bodied Chevrolet is used as logistics support for the cinema camera teams. The Corps insignia was not visible on vehicles deployed overseas. (U.S. Army SC-114476)

SIGNAL CORPS MODEL K VEHICLES

K Model	Organic Definition	Manufacturer	SCR No.
K-18	Truck, Van Operating, 1½-ton, 4×2	General Motors Truck & Coach Division	SCR-197
K-19	Trailer, Van operating, ¾- or 1-ton, 2-wheel	Miller's Trailers & Fleetwood Glider Trailer	SCR-197
K-22	Trailer, Antenna Mount, 8-ton gross, 2-wheel	Kingham Trailer Co. & Couse Laboratories	SCR-270
K-28	Trailer, Antenna Mount, 4-ton, 4-wheel	Fruehauf Trailer Co.	SCR-268
K-28-B-C	Trailer, Antenna Mount, 4-ton, 4-wheel	Fruehauf Trailer Co.	SCR-268
K-29	Trailer, Van Operating, 1-ton, 2-wheel	Adam Black & Son.	SCR-277
K-30	Truck, Van Power, 4-ton, 4×4	General Motors Truck & Coach Division, AFKW 804	SCR-270
K-30-B	Truck, Van Power, 4-ton, 4×4	Autocar Company Model U-8144	SCR-270
K-31	Truck, Van Operating, 4-ton, 4×4	General Motors Truck & Coach Division, AFKX-804	SCR-270
K-31-B	Truck, Van Operating, 4-ton, 4×4	Autocar Company Model U-8144	SCR-270
K-32	Truck, Tractor, 5-ton, 4×4	Autocar Company Model U-5044	SCR-270
K-33	Truck, Cargo, 1½-ton, 4×4	Chevrolet Motor Division Model G-4103 and G-7123	SCR-270
K-34	Trailer, Van, 5-ton, 4-wheel	A.J. Miller & Superior Trailer Mfg. Co.	SCR-268
K-34-B-E	Trailer, Van, 5-ton, 4-wheel	A.J. Miller & Superior Trailer Mfg. Co.	SCR-268
K-35	Trailer, Van operating, 2-ton, 2 or 4-wheel	Checker Cab. Mfg. Co. and Meteor Motor Cab. Co.	SCR-270
K-36	Trailer, Telephone Construction & Pole Hauling, 2-ton, 2-wheel	Various	
K-37	Trailer, Telephone Construction & Cable Hauler, 5-ton, 2-wheel	Various	
K-38	Trailer, Telephone Cable Splicer, ¼-ton, 2-wheel	Various	
K-39	Trailer, Van Power, 4-ton, 4-wheel	York Hoover	SCR-270
K-40	Trailer, Van Operating, 4-ton, 4-wheel	York Hoover	SCR-270
K-42	Truck, Telephone Maintenance w/o winch, 1½-ton, 4×4	Chevrolet Motor Division Model G-4112 and G-7173	
K-43	Truck, Telephone Maintenance with winch, 1½-ton, 4×4	Chevrolet Motor Division Model G-4112 and G-7173	
K-44	Truck, Earth Borer, 1½-ton, 4×4 Model G-4112 and G-7163	Chevrolet Motor Division	
K-45	Trailer, Van Photographic, 1½-ton, 2-wheel	Fleetwood Gilder Trailer	
K-50A	Truck, Telephone Maintenance, Slant Box	Dodge & Chevrolet	
K-50B	Truck, Telephone Maintenance, Square Box	Dodge & Chevrolet	
K-51	Truck, Van Radio, 1½-ton, 4×4	Chevrolet Motor Division Model G-4112 et G-7105	
K-52	Trailer, Power Engine, 1-ton, 2-wheel	Various	
K-53	Truck, Van Multi-Purpose, 2½-ton, 6×6	General Motors Truck & Coach Division, CCKW-353	
K-54	Truck, Cargo, 1½-ton, 4×4	Chevrolet Motor Division Model G-4103 and G-7123	SCR-270
K-55	Trailer, Van, 1½-ton, 2-wheel	A.J. Miller & Oneonta Linn	
K-56	Truck, Van Power or Operating, 6-ton, 6×6	White Motor Co. Model 666 & 666CE	SCR-268 SCR-545
K-57	Truck, Van Power, 2½-ton, 6×6	General Motors Truck & Coach Division, CCKW-353	SCR-527
K-58	Trailer, Antenna Mount, 4-ton, 4-wheel	Fruehauf Trailer Co.	SCR-268
K-59	Truck, Van Multi-Purpose, 2½-ton, 6×6	General Motors Truck & Coach Division, CCKW-353	
K-60	Truck, Van, 2½-ton, 6×6	General Motors Truck & Coach Division, CCKW-353	SCR-268 SCR-527 SCR-545
K-60-B-D	Truck, Van, 2½-ton, 6×6	General Motors Truck & Coach Division, CCKW-353	SCR-268 SCR-527 SCR-545
K-61	Truck, Van Multi-Purpose, 2½-ton, 6×6	General Motors Truck & Coach Division, CCKW-353	
K-62	Truck, Van, 4-ton, 4×4	Autocar Company Model U-8144	SCR-270
K-63	Trailer, Power Engine, 1-ton, 2-wheel	Various	

K-64	Trailer, Antenna Mount, 8-ton gross, 2-wheel	Kingham Trailer Co. & Couse Laboratories	SCR-270
K-64-C	Trailer, Antenna Mount, 8-ton gross, 4-wheel	Kingham Trailer Co. & Couse Laboratories	SCR-270
K-65	Trailer, Van operating, 6-ton gross, 4-wheel	Checker Cab. Mfg. Co. & Meteor Motor Cab. Co.	SCR-270
K-67	Trailer, Antenna Mount, 6-ton gross, 2-wheel	Fruehauf Trailer Co.	SCR-547
K-68	Trailer, Antenna Mount, 8-ton, 4-wheel	Fruehauf Trailer Co.	SCR-268
K-70	Truck, Van, 1½-ton, 4×4	Chevrolet Motor Division Model G-4112 and G-7105	
K-71	Trailer, Antenna Mount, 8-ton gross, 2-wheel	Kingham Trailer Co. & Couse Laboratories	SCR-270
K-72	Trailer, Van, 7-ton gross, 4-wheel	A.J. Miller	SCR-527
K-73	Truck, Tractor, 1.55-ton, 4×4	Chevrolet Motor Division Model G-4112 and G-7113	SCR-547
K-75	Trailer, Antenna Mount, 14-ton gross, 4-wheel	Kingham Trailer Co. and Heil Co.	SCR-545
K-76	Trailer, Turntable Receiver, 5-ton, 4-wheel	Fruehauf Trailer Co.	SCR-527
K-77	Trailer, Turntable Transmitter, 5-ton, 4-wheel	Fruehauf Trailer Co.	SCR-527
K-78	Trailer, Antenna Mount, 12-ton gross, 2-wheel	Fruehauf Trailer Co.	SCR-584
K-83	Dolly, 2-wheel	Fruehauf Trailer Co.	SCR-584
K-84	Trailer, Antenna Mount, 7-ton gross, 4-wheel	Fruehauf Trailer Co.	SCR-784

list, there were 62 models of vehicles and trailers cataloged by the Signal Corps between 1940 and 1943. Most of them were associated with a particular radio or radar installation and were regrouped into series of very specific usages. Throughout the period 1940–5, however, there weren't enough of these vehicles to fulfill all the technical needs, with several other types, which were not referenced, whether by choice or by obligation, used for certain tasks.

The list of Model Ks as above sums up all these vehicles, which are described in the following pages.

First Army maneuvers in August 1939. The 51st Signal Battalion has installed a telephone exchange by hooking it up to a temporary jib. The vehicle, a small 4×2 with a cargo body, has no particular classification. (U.S. Army SC-114358)

Above: The Signal Corps' insignia comprised two signal pennants (since 1868), with a burning torch added in 1884.

Right: Aberdeen Proving Ground. October 12, 1939, was the 20th anniversary of the Ordnance Association. Here, another equipment support vehicle, a 4×2, 3-ton truck belonging to a unit of XXX Army Corps, is being used to transport the sound system for the event. An audience of some 5,000 people, including industrialists and engineers, was present. (U.S. Army SC-114358)

3 Radar Trucks

SCR-268, Mobile Long-Wave Searchlight Control Set

It was in the 1930s that this type of radar was decided on but several more years were needed before a satisfactory resolution was achieved. A lot of tests resulted in successive intermediate equipment before a final, usable design was eventually found.

Although using radio waves as a means of detection was nothing new, it wasn't until 1937/8 that anything concrete took shape, which resulted in direct and indirect technical means for full-scale trials to be carried out with promising results. It was only in 1940 that a more or less compact model came about—the SCR-268 Mobile Long-Wave Searchlight Control Set—whose various antennae were assembled on a single gantry.

This apparatus required three masts, separately in 1938, and then regrouped later: a transmitting antenna, a receiving antenna for defining the azimuth plus the distance from the target object, together with a second receiver to calculate the altitude. The Signal Corps developed this equipment, but it was used by the U.S. Army Air Forces, anti-aircraft artillery units and sometimes even the Navy.

The SCR-268 was a search radar set specially designed for aiming searchlights and AA guns. It worked on the long-wave principle, and it was medium range. Since its design dated back to the mid-1930s, its performance was only approximate but good enough for the searchlight and AA battery servers to aim correctly. It was in service for a long time, right until the end of the war, and was improved regularly. These improvements had an impact on the vehicles and the trailers. At the beginning, two trailers—a K-28 and a K-58—were needed to carry the radar part with two K-56 6-ton, 6×6 trucks and one K-60 2½-ton, 6×6 truck.

Subsequently, one of the two K-56s was replaced by a K-34. Any other vehicle of the 6-ton class or more could tow the trailers provided it had enough power and a compressor for the air brakes.

Right: The elevation antenna enabled the altitude of objects caught in the beam of the transmitting antenna to be determined. Simple trigonometric calculations coupled with the data from the third antenna defined the aiming angles. (U.S. Army Signal Corps)

With the SCR-268's later developments and the creation of a hard-top command post came the K-68 trailer whose payload was almost double.

It was not easy to set up these early radar sets. Once the equipment arrived at its place of operation, it took 13 men and two hours' work to set up. Another four hours were needed to calibrate the electronic elements of the radar.

To improve its locating and identification capabilities, the SCR-268 was generally associated with an RC-148 type IFF (identification friend or foe) set which could detect the beacon in a friendly aircraft.

October 1938. The first of the three antennae making up the SCR-268 is being set up. Called a transmitting antenna, it sent FM pulses toward the horizon. It was positioned on a trailer and had a peripheral deck that prevented electromagnetic reflections from interfering with emissions. (U.S. Army Signal Corps)

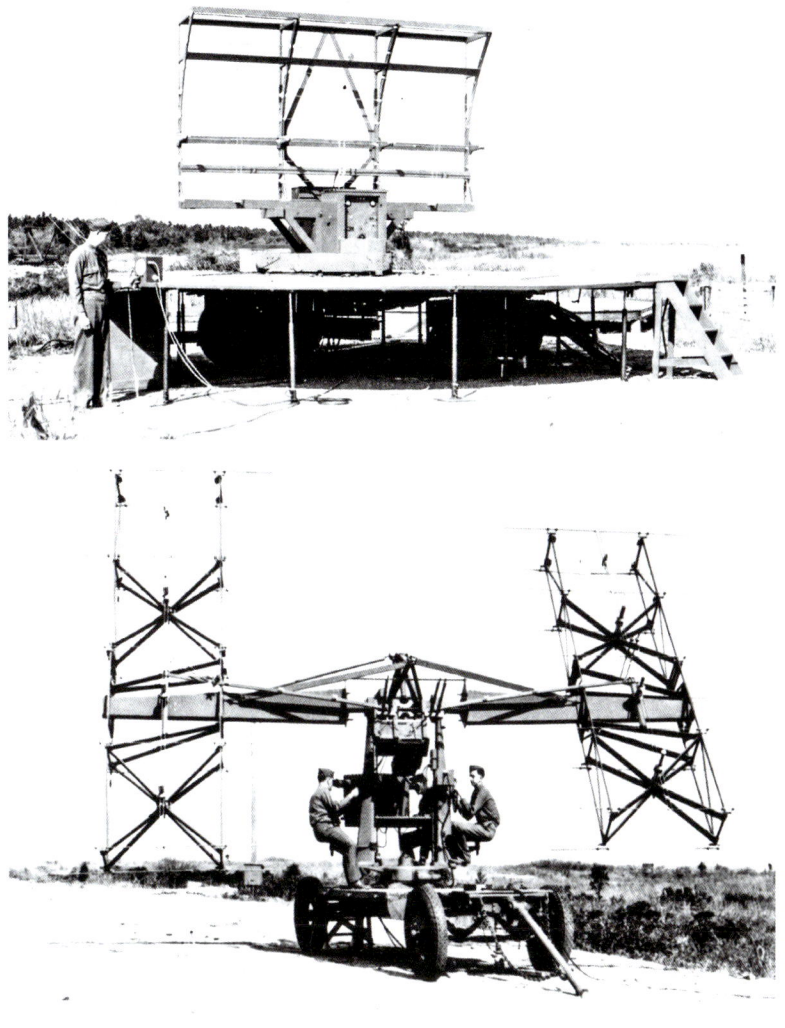

Top: The third set-up, the azimuth antenna, indicated the direction and the speed of the target. (U.S. Army Signal Corps)

Above: Putting the three original antennae together made up the first real SCR-268. Its mission for AA defense meant it did not have to work over long distances; it was effective from 150 feet to 9,000 feet. For the period and the planes in existence at the time, this was adequate. All the adjustment work for the SCR-268 and its antennae was applied to the SCR-270, also being developed during this time. (U.S. Army Signal Corps)

Left Somewhere on an island in the Pacific, an SCR-268 radar is in its standby mode: the antenna panels are in the inoperative position and the operators have momentarily left their workstations. (U.S. Army Air Forces Ref. 69757 A.C.)

K-28, K-58, and K-68 Trailers

The K-28 and K-58 trailers, designated as Trailer, Antenna Mount, 4-ton in military parlance, were platforms designed to take the SCR-268 radar antennae. There were differences between the two models: the former was used to transport elements plus the antenna panels whereas the latter, which bore the SCR-268's pivoting turntable, was fitted with stabilizing outrigger arms.

Built by Fruehauf Trailer Company, they were mounted on two axles with twin 7.50×20 8-ply tires. The tractors were K-56 6×6, 6-ton trucks. These trailers were fitted with a retractable armature which supported a protective tarpaulin.

The K-28 trailer was unique because it existed in three versions. The basic early production K-28 was mounted on 7.50×15 10-ply tires, but with no spare wheel. The K-28B and K-28C had 7.50×20 8-ply tires with a differently shaped covering tarpaulin. Developing the SCR-268 and especially installing a closed-in technical cabin incorporated into the base of the tower, meant that a more solid trailer had to be designed: the K-68 trailer. The payload increased from 10,000 lb to 14,000 lb.

The K-56 Truck

The White Motor Company delivered K-56 van trucks to the Signal Corps, 1,861 examples coming off the production lines in 1942/3. The first 1,326 examples were delivered under the designation Model 666, complete with the technical body installed on the chassis and bearing registration numbers USA-6021082 to USA-6022407. The following 536 examples were listed with the manufacturer as bare chassis, with the Ordnance Department carrying out subsequent assemblies (USA-6044702 to USA-6045233 and USA-6045234 to USA-6045236).

Built by Superior Coach Corporation and Thomas Car Works, the technical bodies had two quite distinct interior layouts. The first (Truck, Van, Power) fed

Top right: Los Negros Island, March 22, 1944. Battery B of the 168th AAA Gun Battalion has installed its SCR-268 in a palm grove. Radar production underwent several modifications, one of the most remarkable being the shape of the lattice beam forming the horizontal support. (U.S. Army SC-257213)

Middle right: The final improvement: a hard-top cabin around the three operators. Two other men were needed for operating the secondary equipment and the generator installed in the truck. (U.S. Army Air Forces Ref. 69747 A.C.)

Right: This recently unloaded K-28 trailer at Ashchurch in England on September 30, 1942 is entirely covered with tarpaulins. (U.S. Army SC-169175-R)

electricity to the SCR-268. A 9 kVA generator supplying 120 volts at 50–60 cycles was installed inside the body, its exhaust pipe coming out through the roof.

The second (Truck, Van, Operating) housed all the radar's control equipment. The two spars made of wood fixed transversally on the roof enabled the hoist that each K-56 was supplied with to be installed. This hoist, once set up on the rear of the truck and equipped with a manual winch, enabled the operators to set up the antennae.

The White 666 was powered by a 6-cylinder in-line, 14,014-cc, water-cooled Hercules HXD engine, rated at 202 bhp at 2,150 rpm. It was fitted with a four forward- and one rear-speed gearbox, and a Timken transfer box. The braking system, also working on the trailer, used compressed air (Westinghouse automotive air brake system).

It was fed by a twin-cylinder compressor driven by the ventilator fanbelts. The compressed air was stored in two cylindrical tanks. Affixed quick hookups for the trailer brakes were situated on either side on the chassis rear. There was also a set of these fixed hookups on the front of the vehicle. The 80-gallon fuel tank was installed on the outside of the left-hand-side chassis beam, just in front of the rear wheels. With a full tank and 10×202 12-ply tires, the 666 had an average range of about 300 miles, and a consumption rate of 3.75 mpg The electrical set-up was standard 6 volts; it carried two batteries for the 12-volt starter by means of a well-thought-out use of the hookup.

K-60 Truck

The K-60 designation was the Signal Corps' generic definition for a closed-in van used for transporting fragile but bulky equipment.

The technical body was mounted on a 2½-ton, 6×6 CCKW-353 chassis from General Motors Truck & Coach Division. These CCKW-353-1s (and -2s without a winch) with a hard-top or tarpaulin soft-top cabin, had a rigid body with several doors and openings.

The technical body was made and assembled mainly by York-Hoover Corporation or Wayne Works which supplied various different side openings, giving the K-60 four subcategories.

Because of the layout, the main objective of which was to provide as much space as possible, the bodywork

Top left: New Caledonia, July 1944. As well as carrying the protective tarpaulins, the K-28 trailer's metal structure was also used to fasten the goods being transported. (U.S. Army Air Forces Ref. 71622 A.C.)

Left: Brooklyn Army Base, New York, August 18, 1941. A K-28 trailer about to be winched onto a barge, then transshipped to a ship bound for Hawaii. (U.S. Army SC-124598)

Left: The White 666 was a powerful 6-ton, 6×6 truck designated K-56 in Signal Corps nomenclature. This recently unloaded example is towing a K-28 trailer. It is accompanied by a K-60 GMC CCKW-353 truck, suspended in the air, and two GMC CCKW-353 Model K-57s. (U.S. Army Signal Corps)

Below: The K-56 was brought out in two models, first as Truck, Van, Power containing the generator powering the SCR-268. This example belongs to C Battery, 184th AAA Gun Battalion. (U.S. Army SC-241014)

Left: A K-56 operating van truck at Caserta, Italy on April 7, 1945. Teletypes and other control and signals equipment have been placed inside the technical body. (U.S. Army SC-208182)

Above: The two wooden beams installed transversally on the roof of the technical body of this K-56 operating van truck were used for the hoist and its manual tackle. These two elements were used for mounting the radar panels on the main mast. (Private collection)

Left: This partial interior view gives an idea of the size of the generator installed in the K-56 power van truck. Above the engine, the square trapdoor in the roof is for the exhaust pipe to be passed through it. (U.S. Army Signal Corps)

In the Signal Corps depots, reserve K models pile up as do the signal repair trucks on GMC CCKW-353 chassis. (U.S. Army Signal Corps)

Left: Italy 1944. The 128th Signal Radio Intelligence Company is using K-60s in pairs to increase the volume of the work area. The lack of windows ensured that there was blackout, with ventilation ensured by an openwork grille in the front, just above the cabin. (Private collection)

Below: The K-60 van truck was a big transport van, totally empty inside. As the Signal Corps' emblematic vehicle, the K-60 became a beast of burden, used in a multitude of roles. (U.S. Army Signal Corps)

Above: Brisbane Australia, December 14, 1943. Among the interior fixtures and fittings of the K-60 is a store of spare parts for an SCR-268, together with workspace. (U.S. Army SC-241039)

Below left: Designated as a Rectificator Van Truck, this K-60 carries all the electronic parts of an RC-148 IFF set. With no space on the GMX

CCKW-353 chassis, the spare wheel was mounted on the rear left-hand inside panel of the body. (U.S. Army SC-241038)

Below right: Another K-60 was needed to carry the external IFF elements. The design of the technical body's wooden floor allowed various supports to be bolted on. (U.S. Army SC-241037)

Above: A K-34 trailer on Makin Island in the Gilbert Islands, November 28, 1943. This type of trailer progressively replaced the K-56 power van truck. (U.S. Army Air Forces Ref. 63233)

Left: With a K-56 truck in the background and in the company of a lot of brass, Major General Spencer B. Akin, Chief Signal Officer, is having the rudiments of the SCR-268 explained to him. The scene took place on December 7, 1943, during an inspection tour in Australia. (U.S. Army Signal Corps)

ROLLING STOCK ASSOCIATED WITH THE SCR-268

	K-28 & K-58 Trailers	K-68 Truck	K-56 Truck	K-60 Truck	K-34 Trailer
Manufacturer	Fruehauf Trailer Co.	Fruehauf Trail. Co	White Motor Co.	General Motors Co.	Divers
Model	–	–	666 & 666CE	CCKW-353	
Orders	–	–	1,861 examples	–	–
Powerplant	–	–	Hercules	GMC	–
Length	194.25 in	207 in	288.25 in	300 in	214.5 in
Width	94 in	94.25 in	96 in	98 in	92.75 in
Height	133.75 in soft top	134.33 in s/top	126 in	130.5 in	121.75 in
Wheelbase	126 in	134.75 in	185 in	165 in	166.85 in
Weight	6,005 lb	7,507 lb	21,220 lb	19,017 lb	18,980 lb
Payload	10,009 lb	14,013 lb	6,177 lb	8,007 lb	10,009 lb
Total weight	16,015 lb	21,519 lb	27,397 lb	27, 025 lb	18,615 lb
Tires	7.50×20 8-ply	7.50×20 10-ply	10.00×22 12-ply	10.00×20 8-ply	7.50×20 8-ply
Top speed	–	–	35 mph	46 mph	–
Fuel capacity	–	–	80 gal	45 gal	–
Range	–	–	300 mi	212 mi	–
Consumption	–	–	3.75 mpg	4.7 mpg	–
Brakes	compressed air	compressed air	hydraulic	hydraulic	compressed air

Los Negros Island with the 168th AAA Gun Battalion, March 25, 1944. Operating the SCR-268 could become uncomfortable depending on the weather conditions. The men were severely put to the test just sitting on a metal shell-shaped seat, open to the elements. (U.S. Army SC-254856)

manufacturer placed a cube-shaped fuel tank on the right-hand chassis beam, and a toolbox on the left-hand one instead of the usual spare wheel, which itself was placed inside the van at the rear of the left-hand side. As there was no room for the battery box on the chassis, this was attached to the running board on the driver's side.

Note that the K-60 was also used for storing and transporting equipment as well as housing cumbersome electric and electronic equipment. It was also regularly transformed into a maintenance workshop. Its bulky volume enabled it to be used for special cases, like a sound mixing studio for army theater shows, or as a Bookmobile mobile library.

These GMCs were designated as 6×6, 2½-ton van trucks and had a 164-inch wheelbase, and were powered by a 4,417-cc, 6-cylinder engine rated at 92 bhp at 2,750 rpm. The engine was coupled to a five forward- and one rear-speed gearbox, together with a two-speed reduction transfer box, 7.50×20 8-ply tires, and a 45-gallon reserve fuel tank. The K-60's average range was 335 miles for a theoretical consumption of 7.5 mpg with a maximum speed of 45 mph.

K-34 Trailer

The K-34 trailer was an example of how military haulage as a whole was rationalized. Why restrict a truck to one single job—transporting a generator—when a trailer could do the same job but cheaper? The K-34 replaced the K-56 power van truck and also the K-57 GMC CCKW-353 truck. Built by A.J. Miller and Superior Trailer Manufacturing Company, the K-34 trailers had a large-sized body placed on a chassis borne by two axles with twin 7.50×20 8-ply tires. Entirely enclosed, it was fitted with two big rear doors, a side door for access to the main control panel and a double door opening on the right-hand side. There was more space than in the truck which already occupied this niche. Various versions, classified from B to E, followed each other right throughout production, with, uniquely, the first examples coming off the production lines having an electric braking system which was later replaced by a compressed-air system.

SCR-270 Early Warning Radar

Although the SCR-268 was the base used to develop the panels of the early radars, the SCR-270 was the testing ground for a lot of the support solutions.

Both the trailers and the trucks were available in various forms, some of which were undocumented and not always openly referenced.

In the range of electronic detection systems, the SCR-270 was an early warning radar. Although it was

Far left: The antenna was the main element of the SCR-270 radar; using a foldable mast which was an integral part of the K-22 trailer, various panels were hooked up, forming the mesh that enabled waves to be transmitted and received. This one was erected on Baker Island, a desert island in the Phoenix archipelago. (U.S. Air Force Ref. 65271 A.C.SC)

Left: The SCR-270 remained in use throughout the conflict. This example was in position on Okinawa in 1945. Despite its impressive size, with a height of about 55 feet, its lightweight, open structure made it very difficult to spot with the naked eye. (U.S. Air Force Ref. 69746A.C.)

K-62 Truck

In 1943, the Signal Corps ordered 65 new examples of the van truck from the Autocar Company, but since the design had been rationalized, there was no longer any question of there being two types of technical body. A single model, still delivered by the York-Hoover Company, did the job of both the K-30 and the K-31. As for the B version of these two, the K-62 was made using the U-8144-T chassis, the tractor truck used for the pontoon trailers with a soft-top cabin configuration. The manufacturer's type lost its T (for Tractor) on the model towing semi-trailers. The real difference apart from the technical part was that the van truck had no winch.

K-39 and K-40 Trailers

These made up the technical support, accompanying the K-64-C semi-trailer for the SCR-270 in its version intended for the Navy.

They were also created because they would be close to where they were going to be used and did not have to move much, and so trucks did not have to be taken

Rendova Islands in the Solomon Islands, in the South Pacific, July 16, 1943. An SCR-270 has been set up under the palm trees. The technical support is an Autocar U-8144, a K-30B power van truck, and a K-31B operating van truck. (U.S. Army Signal Corps)

K-57 Truck

The K-57 power van truck was none other than a K-60 van truck (see the SCR-268, above) in its general form. There was a trapdoor in the roof to allow the dismountable part of a PE-137-A generator's exhaust pipe to stick through when it was installed.

These GMC trucks were designated as 6×6, 2½-ton power van trucks (or officially Truck, Van, Power) made on a 164-inch wheelbase. They were powered by a 4,417-cc, 6-cylinder GMC engine, rated at 92 bhp at 2,750 rpm. It was coupled to a gearbox with five forward and one rear gears, 7.50×20 8-ply tires and a 45-gallon fuel tank. It had a range of about 335 miles with a theoretical consumption rate of 7.5 mpg at a maximum speed of 45 mph.

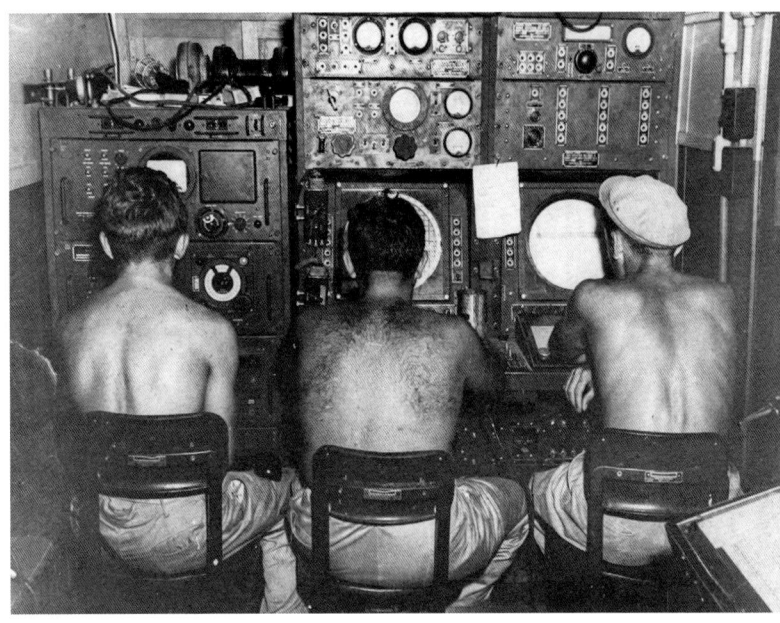

Right: A view of part of the internal technical equipment of the K-72 van trailer. Three men were needed to monitor and analyze the control screens which received the information from the K-77 and K-76 trailers. (U.S. Army Air Forces Ref. 69758 A.C.)

Above: March 1945, Bonin Island, Iwo Jima. Preparing the site for an SCR-527 radar station. The K-76 and K-77 are clearly visible, as is a K-57 truck. The two ditches made with the help of bulldozers were there for digging in the encampment and part of the technical equipment. It was the only way of protecting and camouflaging both men and technical equipment in this desert-like environment. (U.S. Army Air Forces Ref. 64571 A.C.)

Left: Luzon Island, Philippines, January 1945. C Company of the 565th Signal Air Warning Battalion has set up its SCR-527 radar. The threat of an attack being minimal, there is no camouflage—all the equipment is installed in the middle of a field. (U.S. Army Air Forces Ref. 60778 A.C.)

SCR-545 Mobile Long-Wave Aircraft Detector

Entering service in 1943, the SCR-545 was an upgraded extrapolation of the older SCR-268. The antenna, the control post, and all the electronics were assembled in a single trailer to optimize moving and setting up.

The SCR-545 was a radar detection system designed specifically for aiming AAA guns. It worked on the long-wave principle, and it had a long range. Its optimal operating range was between about 1,500 feet and 150,000 feet. It could determine the direction and the altitude of the aircraft it spotted. Its capability was only approximate given the age of the initial technology, but it was enough for aiming the searchlights and guns correctly. It was only used for a short time as technical progress made over the same period, on the SCR-584 for example, very quickly rendered it obsolete, so much so that only 50 or so examples were ordered.

Its transportation was light, made up as it was of the K-75 trailer, a K-56 6-ton, 6×6 truck (like that for the SCR-268), and a K-60 2½-ton, 6×6 truck defined as a spare parts truck. If an RC-145 IFF set was assigned to the grouping, a second K-60 truck was added. One of the advantages of this version was the speed in which it could be set up—less than an hour to get it functioning and only four men were needed to operate it.

K-75 TRAILER SPECIFICATION	
Manufacturer	Kingham Trailer Co.
Model	–
Orders	about 50
Powerplant	–
Length	301 in
Width	101 in
Height	129.5 in folded
Wheelbase	199 in
Weight	14,081 lb
Payload	14,143 lb
Gross weight	28,234 lb
Tires	9.00×20 10-ply
Top Speed	–
Fuel capacity	–
Range	–
Consumption	–
Brakes	compressed air

K-75 Trailer

Designed by Western Electric on a Kingham Trailer Company base, with Heil Company coachwork, the K-75 14-ton antenna mount trailer, with double axles and twin 9.00×20 10-ply tires, was possibly over-engineered. To tow such a trailer, a 6-ton, 6×6 truck like the White K-56 was needed. It also supplied

Above: Taken from the 1943 TM 9-2800 manual, this view shows the K-75 antenna mount trailer. Apart from generating electricity, all that was needed for the radar was packed into this single trailer.

Left: Leyte Island, Philippines. This poorly camouflaged and inadequately protected K-75 trailer has fallen victim to a Japanese attack. The command post and the antenna have been seriously damaged. (U.S. Army Air Forces Ref. 60817 A.C.)

compressed air for the braking system. Although combining all the radar elements under one roof, its height made it vulnerable to attack and it needed much protection.

K-60 Truck

Described technically with the SCR-268, the K-60 van trucks associated with the SCR-545 were, for two of them, unique. The one accompanying the IFF set was fitted out normally; the one used for setting up the radar had the designation spare parts truck, or K-60 Work Truck. It carried all the equipment and the spare parts for long-term missions, as well as chests used as workbenches. It was equipped with a small manual gantry crane attached to the technical body's structure.

The third K-60 able to follow an SCR-545 was defined as a K-60 converter truck; it was fitted out with a powerful current converter which meant it could be connected to the civilian network even at 220/240 volts, and then retransmit a 110-volt current to the K-56 truck which modulated it toward the K-75.

SCR-547 Optical Height-Finder Radar

The SCR-547 was a transition system to complement the old SCR-268s and SCR-270s.

It gave them the angular dimension which they lacked, this triangulation being essential for aiming the AAA batteries.

Above left: The inside of the command post installed in the front of the K-75 trailer. The progress made in minimizing the size of components enabled the space needed for operating the SCR-545 radar to be reduced. (From TM 11-1327-P1)

Above right: In the Lunéville region of eastern France, the 108th AAA Battalion has dug in one of its SCR-545 radars with only the antenna above ground and visible. On the right, the RC-145 IFF set adds to the installation's effectiveness. (U.S. Army SC-233228)

Far left: The K-60 van truck accompanying the SCR-545 radar combined a spare parts store and a repair workshop. The technical operation manual for the SCR-545 radio set illustrates the small gantry crane with a manual hoist fitted to the ceiling of the technical body.

Left: The K-60 in the electric-current converter version was equipped in this way when it came out of the Signal Corps workshops.

The SCR-547 was a hybrid radar system designed to give the fire direction aimers the oblique distance between the radar and its target. By trigonometry and by including the angle of inclination of the parabolas, the altitude and the horizontal distance of the monitored aircraft was thus obtained. The method of functioning—simple enough—depended on the emission of a radio-electric impulse over a very short wavelength (about 4 inches) by one of the two circular reflectors. This impulse was reflected by the target and picked up by the second parabola; the time the return trip took enabled the coordinates of the monitored aircraft to be calculated. With the advent of the SCR-545 and SCR-584 radars, the SCR-547 was soon classified as obsolete. As it was added to installations already well supplied with electricity, its service vehicles were reduced to a K-67 semi-trailer and a K-73 truck.

K-67 Semi-Trailer

Only a few examples of this trailer for limited distribution were built by Fruehauf Trailer Company, with one aim only: to be used for the SCR-547. Built on a profiled chassis on a single axle, it was equipped with two sets of double wheels with 7.50×20 8-ply tires, a jockey wheel, and two side stabilizer arms. Its regular tractor was the K-73 Chevrolet 1½-ton, 4×4 prime mover, which was readily equipped to ensure the trailer's electric braking system operated properly.

K-73 Truck

Belonging to the manufacturer's types G-4112, G-4165, or even G-7113—the three big production series—the K-73s were only a minor part of the 3,530 examples built in this 1½-ton, 4×4 category.

These 1½-ton, 4×4 tractor trucks were built on a 145-inch wheelbase. They were powered by a Chevrolet 3 860-cc, 6-cylinder in-line engine rated at 83 bhp at 3,100 rpm. It had a gearbox with four forward and one rear gears, coupled to a two-speed reduction transfer box; the tires were 7.50×20 8-ply; the fuel tank had a capacity of 30 gallons giving the trucks an average range of 180 miles as soon as they were hitched to a trailer, or a consumption rate of 6 mpg, with a maximum speed of about 48 mph.

K-67 TRAILER AND K-73 TRUCK SPECIFICATIONS

	K-67	K-73
Manufacturer	Fruehauf Trailer Co.	Chevrolet
Model	–	G-4112-4165-7113
Orders	–	3,530 examples
Powerplant	–	Chevrolet
Length	248 in	206 in
Width	92.5 in	86
Height	50.75 in (vide)	87 in
Wheelbase	–	145 in
Weight	6,198 lb	6,134 lb
Payload	5,993 lb	4,495 lb
Gross weight	12,485 lb	10,628 lb
Tires	7.50×20 8-ply	7.50×20 8-ply
Top speed	–	48 mph
Fuel capacity	–	30 gal
Range	–	180 mi
Consumption	–	6 mpg
Brakes	electric	hydraulic

Top: At Haleiwa, Hawaii, an SCR-547 optical height-finder radar in position for surveillance, partnering an SCR-268 mobile long-wave searchlight control set visible behind it. An M1 telemeter added to the precision of the work. (U.S. Army Air Forces Ref. 65278 A.C.)

Above right: The K-73 Chevrolet tractor trucks were straightforward little trucks without any embellishment. This second-generation model was called the G-4165-ZK by its manufacturer. (U.S. Army Signal Corps)

Right: A demonstration of how an SCR-547 radar works, in front of a group of high-ranking officers, at Eagle Far, Brisbane, Australia, December 7, 1943. (U.S. Army Signal Corps)

SCR-584 Anti-Aircraft Fire Control Set

During World War II, the SCR-584 was the most accomplished development in mobile radars.

This was a condensed version of the SCR-268 and the SCR-247, brought together in a single trailer, with a reasonably sized foldable antenna. It was a system

Left: August 4, 1944, Dymchurch, Kent, England. An SCR-584 belonging to the 124th AAA Battalion is in position above a steep, narrow road. Power is provided by the generator on an M7 trailer. Armed with 90-mm AA guns, the unit was tasked with shooting down V-1s. (U.S. Army Air Forces Ref. 62969 A.C.)

Below: Oahu, October 9, 1945, in the 584th Signal Depot Repair Shop. Two brand-new SCR-584s have just been delivered; their trailers are still protected at the door and panel junctures. The semi-trailers are fitted with the K-83 dolly. (U.S. Army SC -218251)

conceived for directing AA gunfire. Though designed in April 1942, it only became operational in 1944. Many innovations were incorporated into it: the parabolic antenna was accurate to a degree not yet attained because of the sweep of its beam and following up identified targets was now no longer manual but automatic. The SCR-584 had a range of 45 miles and was really effective up to 30 miles, its only weakness being following targets flying at low altitudes. Grouped in fives in the AAA battalions, or one per battery and one in reserve, the SCR-584 was housed in a K-78 semi-trailer.

K-78 Semi-Trailer

The K-78 12-ton-gross van semi-trailer was built in the normal way with a side opening with double shutters and adjustable rear panels. Mounted on a twin-wheeled axle, it had no regular K truck partner, since any of the 4- to 5-ton tractor trucks could be used, be they Autocars, Federals, or Internationals. Once again, Fruehauf Trailer Company produced this semi-trailer with pneumatic brakes.

Not all AA batteries were partnered with trucks for towing their materiel. Some had tracked vehicles for this. In this particular case, the Signal Corps ordered an indispensable accessory, equipping the K-78 with a second rolling axle—a dolly designated K-83, which was placed under the front.

K-78 TRAILER AND K-83 DOLLY SPECIFICATIONS

K-78		K-83
Manufacturer	Fruehauf Trailer Co.	Fruehauf Trailer Co.
Length	245.5 in	147 in
Width	97.25 in	99.5 in
Height	125.25 in	56.5 in
Weight	9,489 lb	2,997 lb
Payload	9,988 lb	–
Gross weight	19,478 lb	–
Tires	9.00×20 10-ply	9.00×20 10-ply
Brakes	compressed air	none

Top: Internal photo of the radar double command post located in the K-78 semi-trailer. (U.S. Army Air Forces Ref. 92673 A.C.)

Middle: Setting up an SCR-584 in the Philippines, 1945. Apart from the protective ditch that had to be dug, only half an hour was needed to set up this type of radar, compared with six hours for the old SCR-268. (U.S. Army Air Forces Ref. 60720 A.C.)

Right: Tacloban on Leyte Island, Philippines. "A" Battery of the 168th AAA Battalion has set up its SCR-584 radar-carrying K-78 semi-trailer. In the foreground is the IFF antenna for recognizing friendly aircraft. (U.S. Army SC-254973)

a base that already existed in AAA units. Incidentally, it was the last item to be incorporated into the list of K models, with the number 84.

K-84 Trailer

The K-84 7-ton-gross antenna mount trailer was like all the low-bed trailers that equipped the AAA: it had a low platform mounted on two axles with single 9.00×20 10-ply tires. It was slightly bigger than its stablemates, the M1, M7, M14, M17, M18, and M22 trailers, but with a greater payload and suspension improved by incorporating pneumatic shock absorbers. It was usually allocated to the tracked high-speed tractor category of vehicles, but there was nothing to stop it from being hitched up to a 4-ton, 6×6 truck or even in an emergency a 2½-ton, 6×6 truck.

Top: Hitched up to a high-speed M5A1 tractor, this K-84 trailer is in its road configuration. There is no cabin for the radar operators who have to lift up the rear panel to work. If the battery required it or if weather conditions were bad, a small tent could be attached to the rear of the trailer. (U.S. Army S-347051)

Above: It was in the 1950s that the SCR-784 radar really came into its own, during the Korean War. Here the 18th Field Artillery Observation Battalion deploys in the Ka-san region on August 29, 1950. (U.S. Army SC-347055)

SCR-784 Compact Set

The SCR-784 was not really a new radar, but rather an attempt to make the set as compact as possible and, above all, to increase mobility.

As soon as the SCR-584 was designed, and after a promising start, the Army began demanding that radars be reduced in size. It was vital to have equipment available that was easily transportable, even if it meant a reduction in performance. Transforming the SCR-584 into a compact grouping took time, with results only appearing at the end of the war. However, the SCR-784's capabilities were satisfactory: setting it up only took a few minutes but, above all, it was positioned on

K-84 TRAILER SPECIFICATIONS	
Manufacturer	Fruehauf Trailer Co.
Length	220 in
Width	460 in
Height	128.75 in (tarpaulin)
Wheelbase	48.25 in
Gross weight	13,150 lb
Tires	9.00×20 10-ply
Brakes	electric

4

Radio Trucks

SCR-197 Ground Radio Set

Among the extraordinary array of wireless sets available, in the end there were only three types which had special vehicles and trailers under the Signal Corps' "K" designation.

The SCR-197 was, above all, a complete radio station working with the high frequencies. It had two main users: land forces for long-wave communications, and air forces for transmissions between bases for coordinating large-scale air operations.

Right: Ordered in 1941, the K-18 truck was still made by GMC, but this time with the reference CF-351. It could be distinguished by its headlights, now lodged in the wings. The large protective grille for the radiator was unique because it was articulated to enable the engine hood to be opened. (U.S. Army Air Forces)

Below: An SCR-197 ground radio set ready for work. All the components were installed in the K-18 truck and the K-19 trailer. The GMC truck is an AF-361 which can be recognized by the position of the headlights on the radiator grille structure.

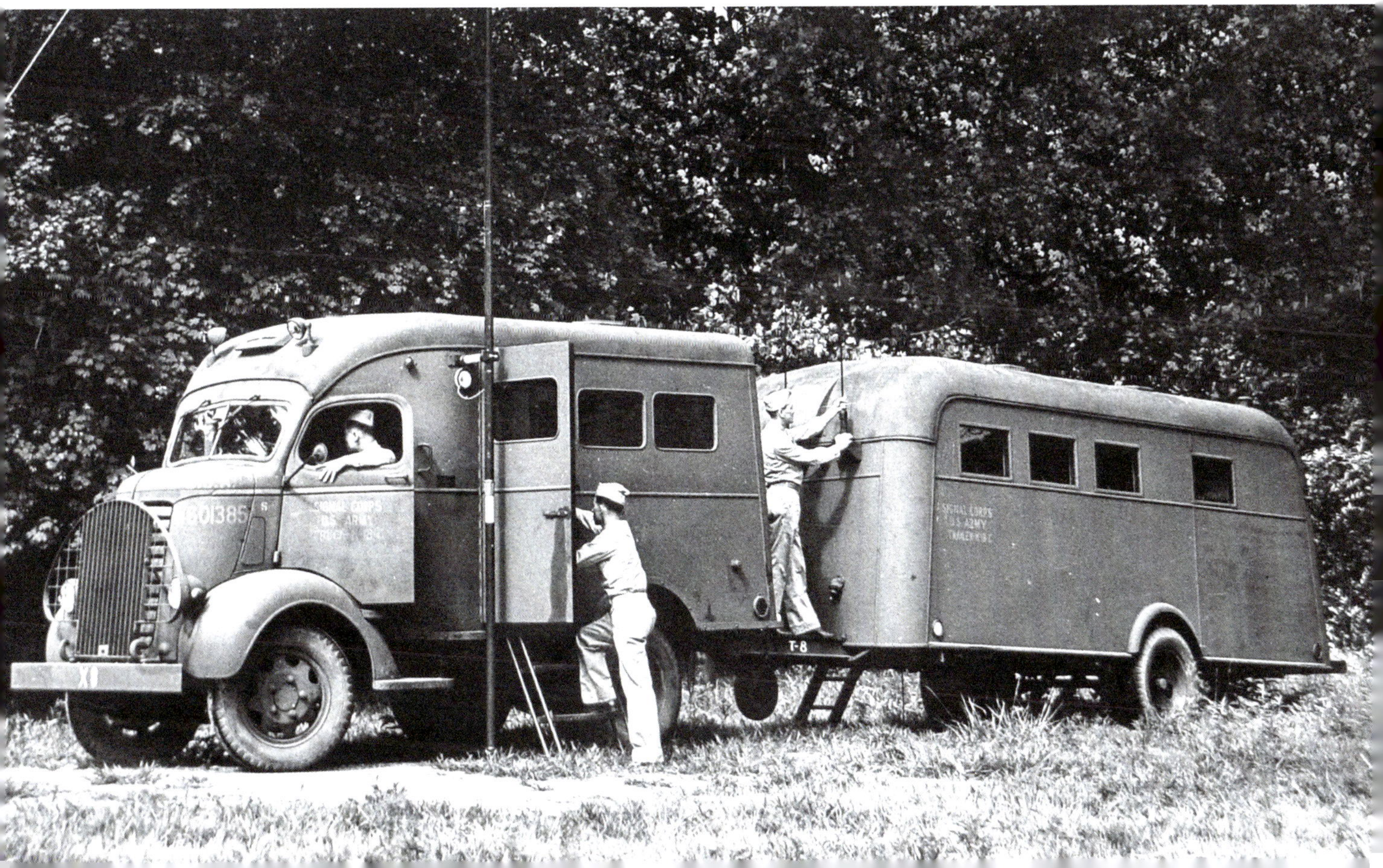

The SCR-197 was designed originally as a static station but by 1938 it proved necessary to make it more autonomous and maneuverable. The most difficult problem was housing the radio's 140 cubic feet of equipment, weighing around 2½ tons, inside a vehicle. Commercial vans were at first bought for just that purpose, but it was only in 1940 that specific military requests were made to the manufacturers. To make the SCR-197 system manageable from a mechanical point of view, the truck and trailer pair was chosen. Under the designation K-18 Truck and K-19 Trailer, the trucks were supplied by General Motors Truck & Coach, and the trailers by Fleetwood Trailers and Miller Trailers.

The SCR-197 radio station had a range of between 100 and 1,000 miles. Apart from the truck and the trailer, it was mainly made up of a BC-325 transmitter and three BC 342 radios for reception. It worked over a frequency spread of 1.5 to 18 MHz, its electricity being supplied by a 220/110-volt 60-cycle GN-42-A generator for transmission and two portable 110-volt 60-cycle PE-75 generators for reception. It needed a 45-foot antenna for transmitting whereas one to three 15-foot antennae were needed for receiving.

Designed mainly for the U.S. Army Air Forces, the SCR-197 was also tested for communications with land units. It accompanied cavalry and infantry units during the Third Army's large-scale maneuvers in May 1940 but results from its equipment were unsatisfactory. At best, the radio exchanges did not range beyond about 40 miles. The frequent moves got the better of the fragile tubes, while the K-18's 4×2 drive did not permit off-road driving.

The war in Europe motivated the Signal Corps technical services to actively look for more effective and especially more mobile equipment. It was critical to solve the problem of the space the radio equipment took up and to make it usable when the vehicle was on the move.

Even if the air forces continued to use the K-18 trucks with the K-19 trailers, their days were numbered: they were classified as obsolete at the same time as the last orders were placed in 1942.

Top: Some 20 or so K-18s were built at the same time as the SCR-197, using the 1938 GMC truck as a base. These trucks had a special protective grille for the radiator. They were also referenced as AF models. (U.S. Army Air Forces)

Middle: Early K-19 trailers were not equipped with the three antenna supports at the front. Signal Corps laboratories, Fort Monmouth. (U.S. Army SC-118498)

Left: These men from the 109th Observation Squadron are clearly happy with one of their K-18 1938 Model trucks. (U.S. Army Signal Corps)

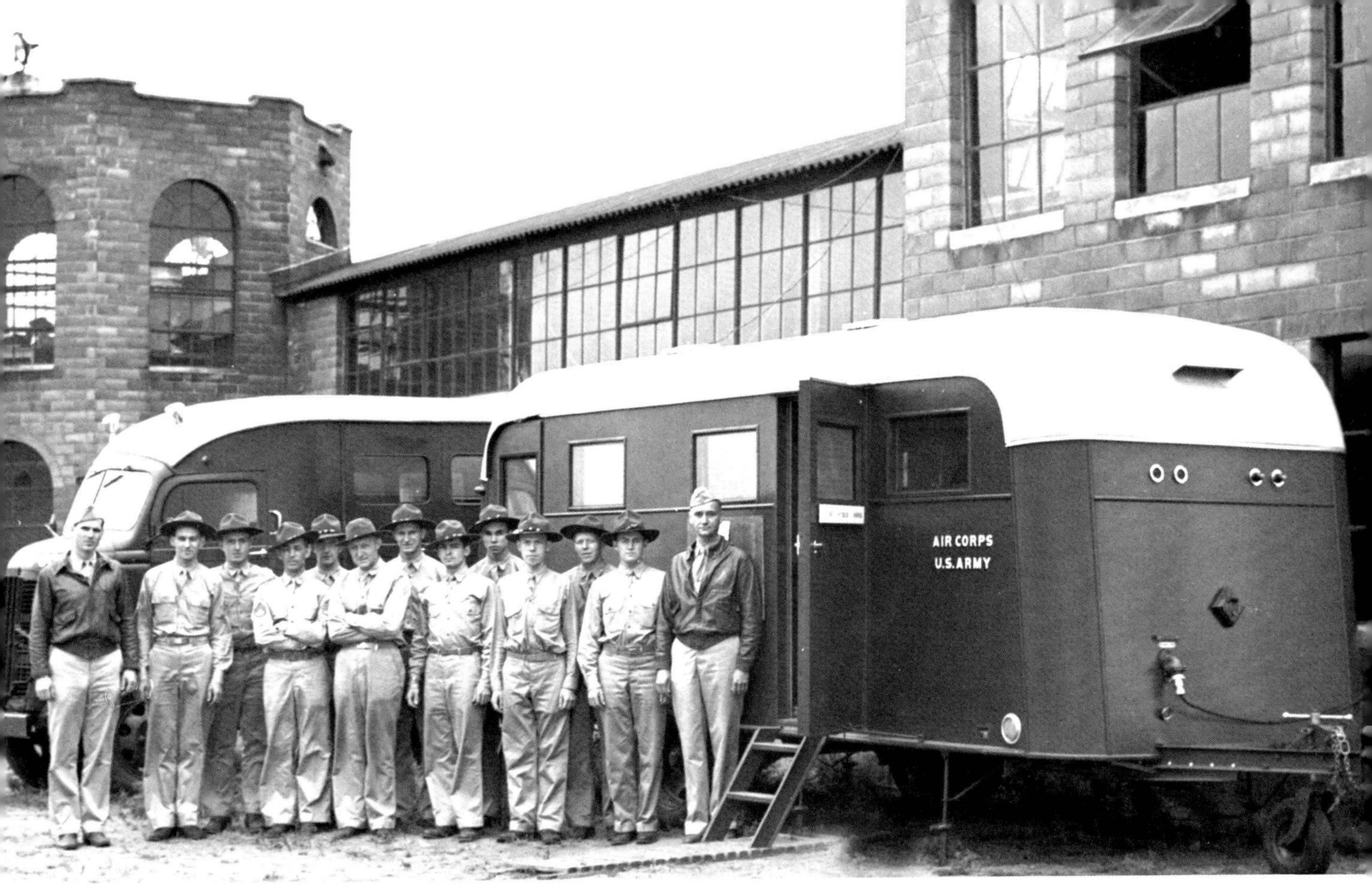

Above: A group photograph in front of a K-18 and K-19 truck and trailer pair, both early types. The roofs were painted white as the trucks were used at aerodromes for pilot identification. (U.S. Army Signal Corps)

Right: A K-18 GMC CF-531 truck has deployed its main aerial and it is the truck itself that keeps it upright. Working behind the front lines, the crew was not worried about camouflage and concealment. (U.S. Army SC-169401)

Below: As a simple proximity radio, the K-19 could work alone without necessarily being accompanied by its tractor vehicle. As long as it was plugged into some form of electricity supply, it worked normally for short-range radio communications. (U.S. Army SC-169396)

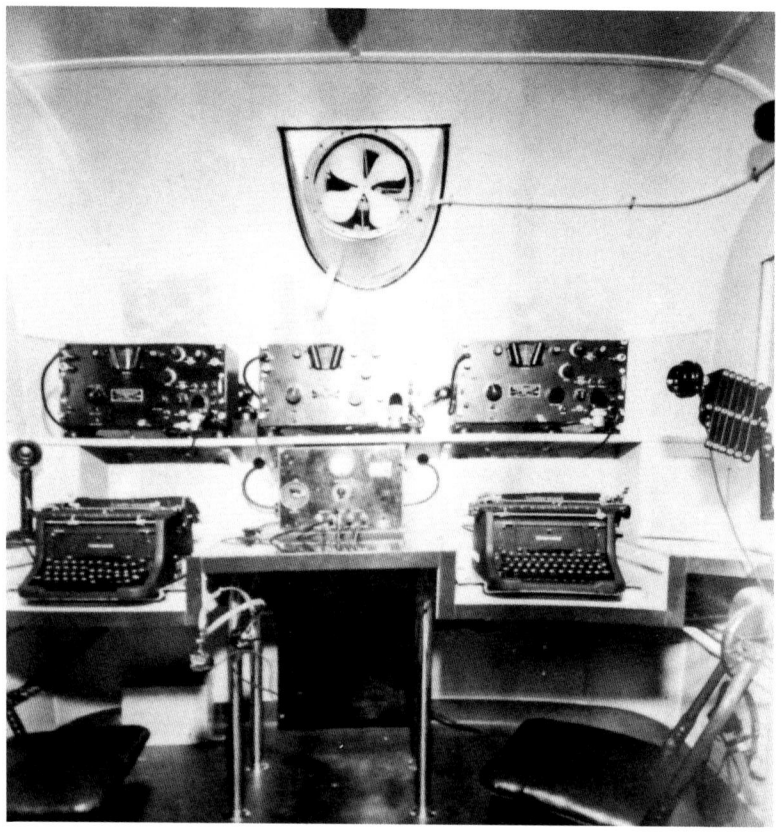

The arrival of the new long-range SCR-299 and its transporter, the Chevrolet K-51 truck, hastened the end of the SCR-197. In spite of this, the USAAF continued to use them, especially in the Pacific theater, where broadcasting on the wavelengths was not troubled by forests and mountains, just the ocean between the airplanes and the operators.

K-18 Truck

Depending on the number of orders and the production years, two models of General Motors Truck and Coach trucks were supplied, receiving the same K-18 designation from the Signal Corps. In 1940, 47 examples of the GMC AF-361 were delivered but they were 4×2, 1½-ton trucks with a 1544 Model cabin and a technical body built by Luce Van Body. Their registration numbers were USA-W-601379 to USA-W-601425. Three points distinguished these early models from the others: their wheelbase was 107.5 inches, their hydraulic brake circuit was based only on a master cylinder coupled to a power cylinder, and their headlights projected from the sides of the bodywork protecting the radiator.

When the military ordered new vehicles in 1942, the GMC Model CF-351 was delivered. The cabin was the 1585 type, with a dashboard comprised of entirely military-type dials. The technical body came from the Hicks Body Company. The wheelbase was lengthened slightly to 109 inches. The braking system included a Hydrovac, and the front headlights were recessed in the wings. These vehicles were still 4×2, 1½-ton trucks and were allocated registration numbers USA-W-607386 to USA-W-607640 for 255 of them, and USA-W-343561 to USA-W-343563 for the remaining three. In 1942 a new order for 109 CF-351s was placed (USA-W-60110711 to USA-W-6010819).

The AFs and the CFs were comparable from a mechanical point of view apart from the braking system. These GMC trucks, defined as 4×2, 1½-ton van trucks were powered by a GMC Model 248, 4,064-cc, 6-cylinder engine rated at 100 bhp at 3,000 rpm. The engine was coupled to a gearbox with four forward and one rear gears; there was no reduction transfer box on the 4×2-wheel drives. Orders for both GMC models amounted to 414 examples of the K-18 van truck.

K-19 Trailer

The K-19 trailer was an adaptation of the civilian caravan, the insides of which were entirely stripped to house part of the SCR-197, together with its operators.

Above and left: The front part of the trailer with its two work stations was used as the radio operating section. The rear part housed the message center section. The crew consisted of eight men. (U.S. Army SC-148317/318)

Holabird Quartermaster Depot, Baltimore, November 1941, the first new-generation K-18 AF-361 truck driving for the first time. Extrapolations of civilian coachwork were initially used to produce this truck for military use. (U.S. Army SC-125942)

K-19 TRAILER AND K-18 TRUCK SPECIFICATIONS

	K-19 Trailer	AF-361 K-18 Truck	CF-351 K-18 Truck
Manufacturer	Fleetwood Gilder Trailer / Miller Trailer Co.	General Motors Co.	General Motors Co.
Orders	484 examples	47 ex.	367 ex.
Powerplant	–	GMC	GMC
Length	255 in	206 in	206 in
Width	84 in	87 in	87 in
Height	103 in	112 in	111 in
Wheelbase	–	108 in	109 in
Weight	5,390 lb	6,519 lb	6,556 lb
Payload	2,001 lb	3,003 lb	3,003 lb
Gross weight	7,392 lb	9,522 lb	9,559 lb
Tires	7.00×20 8-ply	7.00×20 8-ply	7.00×20 8-ply
Max. speed	–	50 mph	50 mph
Fuel capacity	–	20 gal	20 gal
Range	–	220 mi	220 mi
Consumption	–	3.2 mpg	3.2 mpg
Brakes	electric	hydraulic	hydraulic

The first 47 examples were supplied by the Fleetwood Glider Trailer Company and were allocated registration numbers USA-W-04606 to USA-W-04652.

These were trailers in the ¾-ton, single-axle category. In 1941, 258 examples were ordered from Miller's Trailer Company but these became 1-tonners and

Inside the K-18 technical body which is entirely taken up with radio equipment. At the front of the body, the main transmitter stands upright against the driving cabin, preventing communication between the two parts of the vehicle. (U.S. Army SC-148319/320)

With the SCR-197 becoming obsolescent so swiftly, most trucks and trailers were left without work even though they were not that old. Although some of them had had full careers, others were reassigned to quite divergent tasks. The K-18 and K-19 pair was transformed into a mobile recruitment office for the Air Force. The truck was used as an office and the trailer an exhibition room. (U.S. Army Air Forces)

were taken back in two series with registration numbers USA-W-017539 to USA-W-017793 and USA-W-046584 to USA-W-046586. In 1942, 109 new examples were bought (USA-W-165745 to USA-W-065853). The final orders, placed in 1943, increased the K-19 trailer pool by 70 new examples (USA-096892 to USA-09941 and USA-032133 to USA-032152). Administratively, the Signal Corps had 484 examples of this trailer at its disposal.

The K-19s could be considered as simple shelters which could just as easily have been a second K-18 truck. In the trailer there were two parts, a radio operating section with two work posts at the front and a message center section at the rear requiring six operators. A small number of K-19s were also used as photographic laboratories because the early disappearance of the K-45 trailers left a gap that some units only filled later with K-70 trucks.

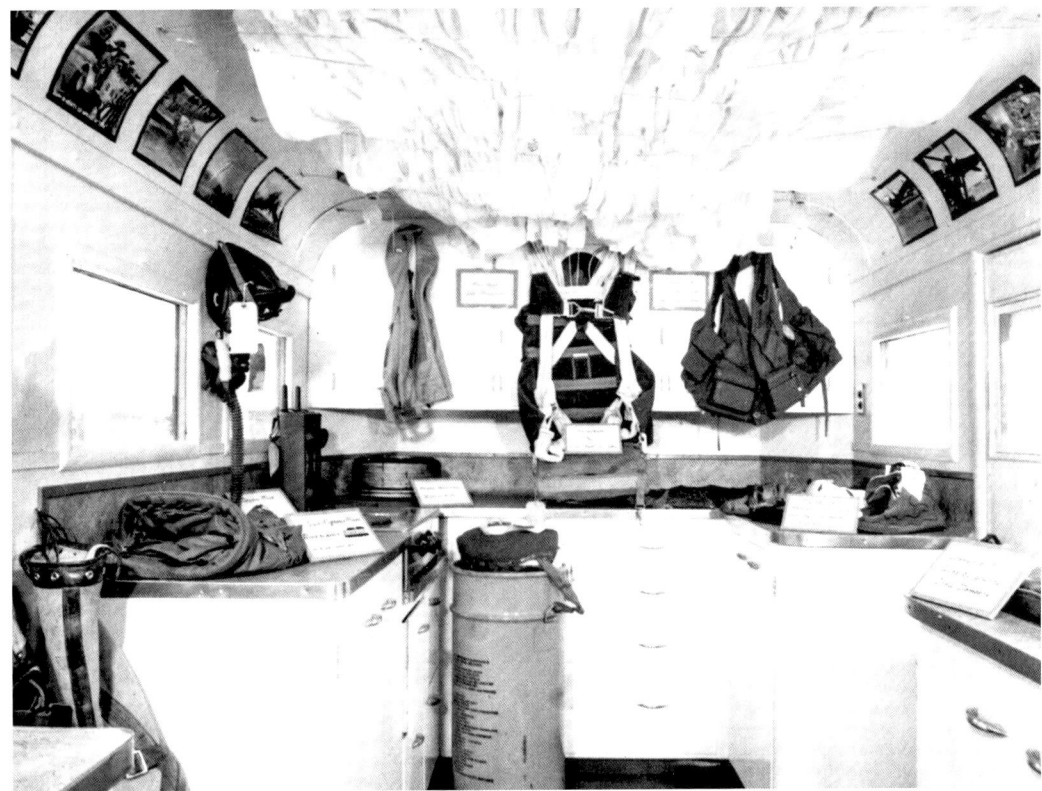

This old K-19 trailer has become a showcase for flight crews. It has been entirely reorganized with new furnishings, sets of drawers, and equipment set up on the work surfaces. (U.S. Army Air Forces)

SCR-277 Radio Range Set

Below: Set up for work, the K-29 trailer was kept horizontal by four props. As the trailer could be towed by different types of vehicles, its coupling was of the variable type. (U.S. Army SC-124614)

Bottom: With the ventilation panels open and the antennae deployed, the K-29's workspace was pleasant for its operators, even in the middle of nowhere. (U.S. Army SC-148316)

At the beginning of the 1940s, the technological developments of radio equipment together with those of the air forces gave rise to the first direction-finding beacon.

The SCR-277, which was still in its infancy, enabled long-range aircraft to easily find their home base after an operation. It was a navigational aid beacon sending radio-guidance signals toward aircraft without radio compasses. Consisting of a BC-467 transmitter with an output power of 800 watts, a BC-468 goniometer, and a BC342 receiver, the beacon functioned on frequencies of 200 to 400 kHz when transmitting and 1.5 to 18 MHz when receiving; electrical power was provided by a single PE-90 generator. Set out in strategic places,

this type of beacon formed an air navigation guidance network. When operational, the transmitter sent Morse signals for the letters A or N in each of the four quadrants around the antenna. Because the signals overlapped, they gave the pilot an idea of his position compared with the beacon. If he was heading for the beacon, he would receive the signal A; if he was going away from it, he would receive the signal N. Navigating from beacon to beacon, a friendly aircraft could therewith make its own way and leapfrog from one beacon to the next. On terra firma, the SCR-277 had an average range of 300 miles. Set up on the heights on an island, this range could reach around 1,000 miles across the water.

K-29 Trailer

Intended only for the air forces, the K-29 trailer, built by Adam Black & Sons, was only a shelter for the SCR-277 and its operators. It was made on a 1-ton chassis, resting on a single two-wheel axle. Built at the beginning of production using smooth sheet metal, it was soon given a casing made of corrugated-iron panels to make it more rigid.

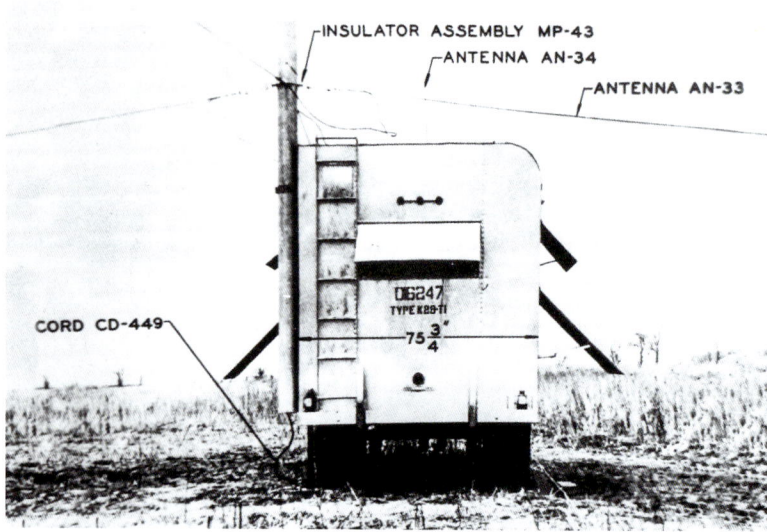

K-29 TRAILER SPECIFICATIONS	
Manufacturer	Adam Black & Sons
Orders	?
Length	172 in
Width	76 in
Height	103 in
Wheelbase	–
Weight	2,284 lb
Payload	2,002 lb
Gross weight	4,286 lb
Tires	7.50×20 8-ply
Brakes	none

SCR-299 Ground Radio Set

Technological innovations at the beginning of the war came in quick succession. The "old" S-197s were replaced as quickly as possible by the new SCR-299s. This meant that a suitable vehicle had to be designed: the K-41 truck and its K-51 trailer.

Comprising a BC-610 transmitter, BC-312 and BC-342 receivers, a BC-614 amplifier, a BC-729 regulator, and a BC-211 frequency meter, the SCR-299 usually

Opposite: Saipan, Mariana Islands, 1945. This mobile range station has been positioned far from anywhere, with the operators getting to work using a service vehicle, here a Dodge WC 51. The definitive version of the K-29 had corrugated metal as its structure, thus making it more rigid. (U.S. Army Air Forces Ref. 65262 A.C.)

The Chevrolet panel truck was the ideal vehicle for carrying the SCR-299 ground radio set components, even though the interior volume was on the small side. This disadvantage was made up for by the addition of side cupboards, one on either side of the vehicle. (Private collection)

worked on a range of frequencies spread over 2 to 8 MHz, but this could be extended to 1 to 18 MHz using conversion kits, its theoretical range being 100 miles; but with favorable weather conditions in the Pacific, certain communications could be extended to 2,300 miles. Its components together with the accessories were installed in four subgroups, all fitted aboard trucks in the manufacturer's workshops at Hallcrafters Company Inc. The large transmitter located on the

mount bracket was affixed to the floor against the driver's cabin. A worktable receiving all the other preassembled radio elements was placed lengthwise on the left-hand side of the body; it was entirely pre-cabled with a work lamp, plugs, and small electric heater.

A wall cupboard was fitted to the right-hand side, for storing various elements like the tuning units, the coil units, and spare batteries. The fourth subgroup was a bench/trunk on which the operators could sit,

Jebinina, Tunisia, May 9, 1943. The electric coupling between the PE-95 generator and the truck is quite visible here, as is the spare wheel's homemade position under the trailer. (U.S. Army SC-380488)

and which housed smaller equipment like radio lamps, quartz, antennae, and bases. The electricity supply for the radio equipment and other elements was provided by a PE-95 generator. This was appreciably the same as the Willys Jeep and provided power of

5 to 10 kilowatts depending on its configuration, for 110–240-volt AC current.

The K-51s and K-52s were served by Signal units which had the challenging task of keeping the larger units in touch with each other for overall coordination.

A K-51 radio vehicle and its K-52 trailer at the Amphibious Training Center, Camp Gordon Johnston, Florida, February 23, 1943. (U.S. Army SC-257527)

In 1942, the 102nd Infantry Division was based at Camp Maxey, Texas. Here, a divisional signal company K-51 is on maneuvers. The inside of the panel truck is just big enough for the radio equipment and two operators. (Private collection)

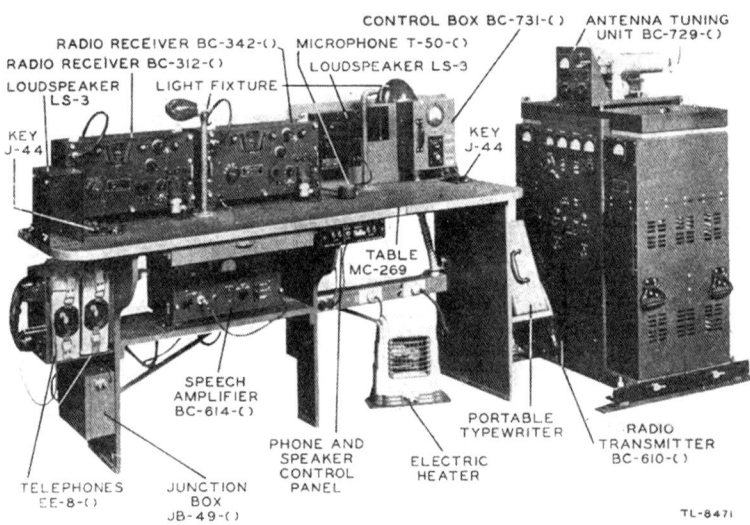

RADIO RECEIVER BC-342-( )   CONTROL BOX BC-731-( )   ANTENNA TUNING UNIT BC-729-( )
RADIO RECEIVER BC-312-( )   MICROPHONE T-50-( )
LOUDSPEAKER LS-3   LIGHT FIXTURE   LOUDSPEAKER LS-3
KEY J-44   KEY J-44
TABLE MC-269
SPEECH AMPLIFIER BC-614-( )
PORTABLE TYPEWRITER   RADIO TRANSMITTER BC-610-( )
PHONE AND SPEAKER CONTROL PANEL   ELECTRIC HEATER
TELEPHONES EE-8-( )   JUNCTION BOX JB-49-( )
TL-8471

Above: This extract from the 1943 technical manual shows similar internal equipment on which several minor improvements can be made out. On the left, under the table, two EE-8 field telephones are visible. (U.S. Army Signal Corps)

Left: Hallicrafters supplied all the radio equipment for the K-51s. This is a full-page advertisement taken from *Radio News* in November 1942.

Below: These two views show how the inside of the K-51 was taken over by the SCR-299 radio equipment. The four blocks are clearly visible: the overhanging cupboard on the right; the big central storage chest, also used as a seat for the operators; at the back the main transmitter; and on the left-hand side, all the communications equipment installed on a pre-cabled table. (U.S. Army SC-174435 & -257528)

As well as the radio equipment for their job as a message center, they could also connect to the telephone lines thanks to two side coils and two field EE-8 telephones under the work surfaces, which also enabled them, at a distance, to control the apparatus, up to a mile and a half away. When this equipment was fitted in the Chevrolet panel truck, long-distance communications became much easier. It was in North Africa, during the Operation *Torch* landings in November 1942, that the K-51s first appeared. The 829th Signal Service Battalion had the difficult task of coordinating the various components of this first major operation by the Western Allies through the main command post set up in Gibraltar.

All throughout the North African Campaign, the K-51s accompanied the advancing troops and provided such a boost to military communications that General Dwight Eisenhower praised their vital contribution to the success of *Torch*. Subsequently they rendered the same service in Sicily and Italy and, during the Normandy landings. These were the only fast method that war correspondents had of transmitting their reports to their editors in England. In the Pacific theater, the Marine Corps used its K-51s right up to Japan, but the days of the Chevy radio panel truck were numbered.

It was too high off the ground and took up too much space aboard ship. In November 1942, a study was made for designing a "box" that could be installed on various chassis, without needing a specifically dedicated vehicle. This was the beginning of the Type HO-17 shelter. The SCR-299 survived the demise

Side view of a G-7105-NG delivery panel truck in one of the ubiquitous Signal Corps depots. The vehicle has not yet been delivered to its unit; it bears no markings except for the factory-painted registration number. (U.S. Army Signal Corps)

Below: Maneuvers in the United States using all available radio means. The truck is hitched up to a K-52 trailer which carries the generator providing all the electric power. (U.S. Army SC-42-MP-1374)

of the K-51, but in a more evolved and specific form for use with the shelter-type bodies; it became the SCR-399 and in its simplified airborne version, it was designated SCR-499.

K-51 RADIO TRUCK SPECIFICATIONS	
Manufacturer	Chevrolet
Model	G-7105-NG
Orders	not known
Powerplant	Chevrolet
Length	221.5 in
Width	87 in
Height	95 in
Wheelbase	145 in
Weight	6,766 lb
Payload	4,030 lb
Gross weight	10,796 lb
Tires	7.50×20 8-ply
Max. speed	48 mph
Fuel capacity	30 gal
Range	180 mi
Consumption	6 mpg
Brakes	hydraulic

K-51 Truck

Orders amounted to 4,034 examples of these Chevrolets hard-top trucks with a 145-inch wheelbase, an unknown number of which were supplied directly to the Signal Corps.

Although the panel version was referenced with the manufacturer under three distinct types, only the third development, the G-7105-NG, was used as a radio vehicle.

Hampton Roads Port of Embarkation, Newport News, May 1943. Among the countless vehicles about to leave for overseas service, the K-51 Chevrolets wait to be loaded. (U.S. Army Signal Corps)

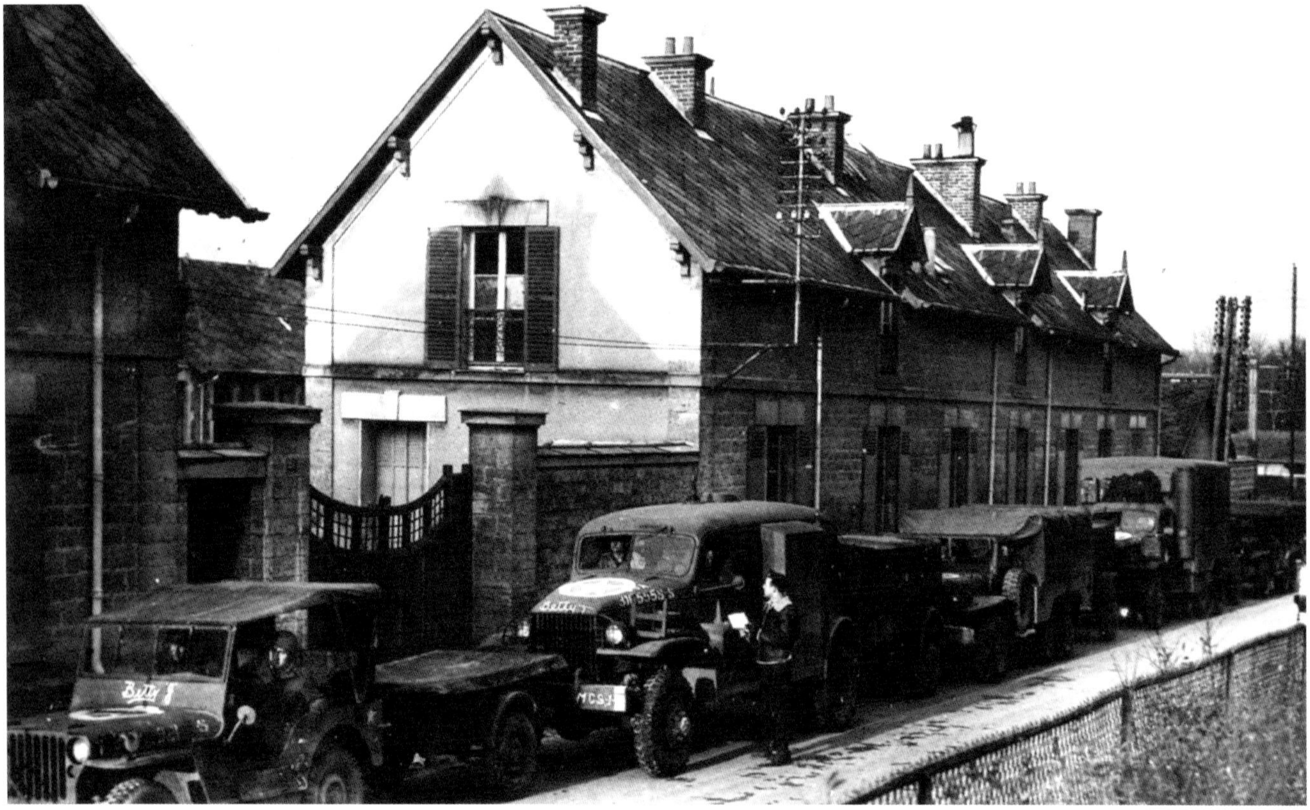

Germany, 1945, the 40th Mobile Command Unit forms up its column to continue its advance into enemy territory. A rare particularity: each vehicle has the same first name followed by a letter—one can make out "Betty G" on the Jeep and "Betty F" on the K-51 Chevrolet, most certainly markings defining the order of the vehicles, added to the regimental markings on the fenders. (Private collection)

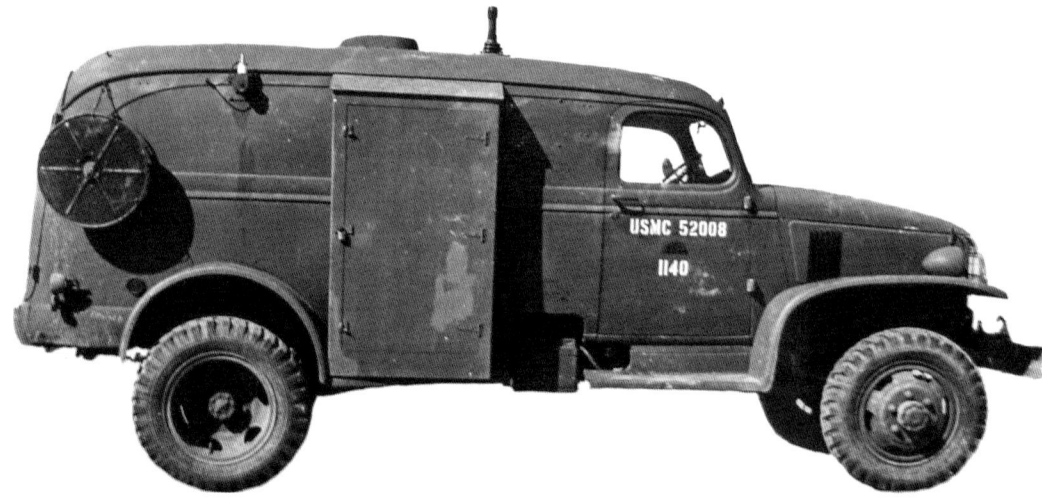

The radio panel truck therefore had the same mechanical features as the other 1½-ton Chevrolets. In the end, the K-51 design was a three-sided design: Chevrolet for building the vehicle, the Signal Corps for adapting it, and Hallicrafters for supplying and assembling the radio equipment.

A cupboard was installed on each outside flank, together with an antenna contact and a large coil of telephone cable on a reel; the roof was perforated for an air circulation fan and was fitted with a base plate for the main antenna, connectable from the inside. Two headrests were fitted for the crew between the driver's compartment and the rear bulwark. Once modified, the vehicles were sent to Hallicrafters, where they were fitted out with the radio equipment.

This truck used a Chevrolet 3,860-cc, 6-cylinder engine rated at 83 bhp at 3,100 rpm. It was coupled to a gearbox with four forward and one rear gears and a two-speed reduction transfer box.

Right: Vehicles identified as Model Ks were not only intended for the Army and the Air Force; the Navy and the Marine Corps also used them. (U.S. Army Signal Corps)

5 Multi-Purpose Vehicles

Multi-Purpose Van

The Van, Multi-Purpose K-53 or rather the K-53 multi-purpose van, on a GMC CCKW-353 base was quite unique, both in the way it was designed and then in the way it was built. It took advantage of all the features thought up by the Signal Corps and combined them into a single technical body.

In 1941, the Signal Corps began studies on several trucks with a rigid technical body to transport and protect special equipment. It was only in 1942 that the first prototypes appeared, among which was a van to shelter the future mobile weather station instruments. Curiously enough, a preproduction-series vehicle, designated CCKW-353-F1, appeared in the GMC registers, with the registration USA-W-434762.

Because of its general shape, it was what the U.S. Army Air Forces requested as its future radio vehicle, later adopted by the Signal Corps with the reference Model K-53. But this unique example was a poorly designed vehicle: mounting the technical body

Above: The first mobile weather station was made using a modified CCKW-353 cargo truck as a base. The lower edges of the technical body were horizontal like the workshop truck, the famous ordnance maintenance trucks of the U.S. Army. (U.S. Army SC-138457)

Left: (PSD file) The GMC CCKW-353-F1 truck was one of a kind because of its shape, and, above all, because of its fuel tank. Justifiably so, as it can be considered the prototype preceding the creation of the K-53 multi-purpose van. (U.S. Army SC-134758)

required the original fuel tank to be moved lower, which reduced its ground clearance and thereby diminished the vehicle's off-road capabilities. This explains why the design was abandoned and restarted using a different base.

K-53 Multi-Purpose Van

The designation K-53 in the Signal Corps designated a multi-purpose van specially dedicated to the air forces. The prime aim of these trucks was to provide a practical shelter for the mobile radio equipment operating on forward airfields where there was no hard-standing infrastructure to take electronic equipment. Because they stayed on site for a long time, it was vital they used rigid technical bodies. Built on the GMC CCKW-353-1 or -2 trucks with or without a winch, they had either an enclosed (hard-top) cabin or a tarpaulin-covered (soft-top) cabin. A small fuel tank was at first located on the outside of the two main chassis spars. The spare wheel, normally on the left-hand side of the vehicle, was repositioned at the rear of the technical body.

When K-53 production was in full swing, a new single fuel tank was fitted to the left-hand spar. The technical body was a cube built around a metal structure, covered with sheet metal on the outside and plywood panels on the inside, the gap between them filled with insulation. Two windows on either side gave natural light and ventilation. There was a

Top: Although the general lines of the body evolved, the rear was already well designed with a wide step and a special mounting for the spare wheel. (U.S. Army SC-138459)

Left and below: The first example in a long series, the final total of which is unknown, the K-53 multi-purpose van was a simple model, without a heating tower or a reinforced roof for setting up antennae. (U.S. Army SC-133815/17/18)

6

Specialized Vehicles

Light Telephone Maintenance and Repair Trucks

This category, comprising the K-50-As and the K-50-Bs, was made up of nine different vehicles built by two manufacturers in two similar forms. Some 1,096 of these vehicles arrived at the Signal Corps between 1940 and 1942.

These small vehicles were perfectly suited to maintaining, repairing, and fitting short telephone extensions to existing telephone installations, which was the Signal Corps' job. The way the bases and the army camps throughout the country developed so quickly at the beginning of World War II meant there was an increase in the number of operations maintaining networks in good working order. These were not vehicles with any martial purpose although a large number of them were four-wheel drive and most of them remained on United States soil. Two versions of the technical body were ordered, known as the Slant Box and the Square Box.

However, orders for these specific vehicles stopped quickly because they cost too much, were too

Above: A similar vehicle but this time the 3103-AK Model, still with the square box-type technical body. The two upper support levels are clearly visible here. (JQD 276-2-41)

Below: Chevrolet Model KC ½-ton, 4×2 truck, on the Pearl Harbor base, 1943. The "U.S. Army" on the driver's door is so the vehicle would not be confused with a Navy vehicle. (Private collection)

U.S. ARMY

accessible from the rear which was nonetheless reduced by the presence of two chests accessible from the right-hand side. The first, lengthwise above the rear wheel, was taken up by three large drawers for storing smaller accessories; the second, vertical, housed various boxes and platforms for housing the equipment. The left-hand flank of the structure had two large supports for transporting a double extension ladder.

Dodge and Chevrolet shared the orders for these small repair vehicles, 959 of them being delivered between 1940 and 1942, including four examples of the Model 3102-BK ½-ton, 4×2 trucks registered as 006313 to 006315, and 006006; and 34 Model 3605-ALs, but this time ¾-ton, 4×2 trucks registered 001392 to 001425—38 Chevrolet vehicles in all.

Dodge delivered most of the orders, or 921 examples, split between four models: 372 vehicles of the ½-ton, or 4×4 type, one T-112 type WC-39 (004339), one T-112 Type WC-50 (006930), and 370 T-215 Type WC-43s (007539 to 007908), plus 549 ¾-ton 4×4 T-214 Type WC-59s (0015366 to 0015914).

cumbersome, and used up a lot of raw materials, like the steel the technical bodies were made of. But as the overseas campaigns developed, so demand increased. The repair units had to fend for themselves and use their imagination in adapting the vehicles they were issued for their maintenance tasks.

K-50-A Slant Box

With the particular features of its closed-in body built by Highway Trailer Company, the K-50-A was used mainly for maintaining and repairing existing telephone installations, not setting up new lines. The technical part comprised a large storage space

K-50-B Square Box

The K-50-B had an open body to allow more voluminous equipment to be carried, like telephone cable reels. Each side panel had different boxes and drawers.

The upper part of the central section was taken up by a double transport rack, the double extension ladder being placed along the top, whilst other lengthy elements were stored under it. Although this version was dedicated to setting up new lines, nothing prevented it from being used for simple repairs.

SIGNAL CORPS ½-TON VANS					
	½-ton	½-ton	½-ton	¾-ton	¾-ton
Chevrolet	**Dodge**	**Chevrolet**	**WC-59**	**WC-61**	
Slant Box	**Slant Box**	**Square Box**	**Slant Box**	**Square Box**	
Manufacturer	Chevrolet	Dodge	Chevrolet	Dodge	Dodge
Model	BK & AL	T-212 & 215	KC & AK	T-214	T-214
Orders	38 examples	372 ex.	79 ex.	549 ex.	58 ex.
Powerplant	Chevrolet	Chrysler	Chevrolet	Chrysler	Chrysler
Length	196 in	194 in	196 in	191.5 in	191.5 in
Width	77.75 in	72.5 in	77.75 in	77.25 in	77.5 in
Height	75 in	81 in	80.5 in	80.5 in	85.75 in
Wheelbase	75.5 in	116 in	75.5 in	121 in	121 in
Weight	4,154 lb	4,154 lb	4,308 lb	5,254 lb	5,404 lb
Payload	500 lb	1,000 lb	500 lb	1,300 lb	1,300 lb
Gross weight	4,654 lb	5,154 lb	4,808 lb	6,554 lb	6,704 lb
Tires	6.00×16 6-ply	7.50×16 8-ply	6.00×16 6 -ply	9.00×16 8-ply	9.00×16 8-ply
Max. speed	49 mph	50 mph	49 mph	55 mph	54 mph
Fuel capacity	15 gal	25 gal	15 gal	30 gal	30 gal
Range	176 mi	190 mi	176 mi	300 mi	240 mi
Consumption	11 mpg	7.5 mpg	11 mpg	10 mpg	10 mpg
Brakes	hydraulic	hydraulic	hydraulic	hydraulic	hydraulic

With time, the complexity and the extent of the telephone installations led to adaptations in the field. Note the position of a bench on the top of the technical body on this K-50-A. It made the GI who controlled the reel relatively more comfortable while operating the RL-31 reel unit. (U.S. Army Signal Corps)

The 176th Signal Repair Company, Townsville, Australia, May 1943. Not included in the issue, the RL-31 reel unit is only lightly fastened in place. (U.S. Army Signal Corps)

The 5th Signal Detachment, Fort Bragg, November 1941. Privates A. M. Johnson and Ralph E. Himes finish off coupling the telephone lines to new barracks. Their vehicle is a Chevrolet K-50-B square box truck. (U.S. Army SC-126365)

Above: Dodge supplied 549 examples of this T-214, 4×4, ¾-ton truck known in-company as the WC-59 and designated as K-50-A slant box, slant because the rear was slanted. (Chrysler Corporation)

Below: Very few Dodge WC-61 trucks were ordered by the Signal Corps, only 58 under the designation K-50-B square box. (Chrysler Corporation)

Left: The light telephone maintenance and repair K-50 used the Dodge ¾-ton, 4×4 Model WC-59 as a base. The vehicle is at Camp Chaffee as shown by the markings on the fenders. On the body side, is the VII Service Command insignia. This graphic representation enables the photograph to be dated after March 24, 1944. (Private collection)

Right: Whether the K-50-A or K-50-B as here, all the light telephone maintenance and repair trucks were unique in that they had a large number of cupboards and drawers containing a multitude of consumables to enable operators to carry out a wide variety of repairs. (Private collection)

The K-50-Bs were rarer, since Chevrolet only supplied 79 of the ½-ton, 4×2 version, or 33 Model KCs (registered from 00916 to 00948), and 46 Model 3103-AKs (003628 to 003672 plus 0027967). As for Dodge, it supplied 58 examples of its ¾-ton, 4×4, T-214 Type WC-31 version with registrations 0051312 to 0051389.

M30 Signal Corps General Repair Truck

The specialist workshop vehicles were another particularity of the U.S. Army, which were shared out among the various services, including the Signal

Bristol, England, January 27, 1943, this Dodge ½-ton, 4×4, Type WC-43, Model K-50 slant box truck belongs to the 56th Signal Battalion, V Corps. (U.S. Army SC-126365)

Corps. Even without the "K" definition, the signal repair trucks were indispensable support vehicles.

A total of 12,342 examples of the basic version of the workshop truck (Ordnance Maintenance Truck)— the emblematic vehicle from the GMC CCKW-353 2½-ton, 6×6 range of trucks—were ordered with 39 different interior layouts. According to the report *Engineering of Transport Vehicles, Chief of Ordnance, Detroit 1942–1945*, 1,037 units were supplied under the appellation Signal Corps General Repair Truck Model M30. They were all fitted with the second model of technical body (Type ST-6 body).

Mixing reaction vehicles was common in the smaller units. Assembled here under a camouflage net are a Dodge WC cargo ½-ton, 4×4 truck, a Diamond T Model 614 2½-ton, 4×2 telephone maintenance truck, and a K-50-A Dodge WC-59 4×4, ¾-ton light telephone maintenance and repair truck. (Private collection)

Whether the cabin was either a hard- or soft-top, the CCKW Signal Corps general repair trucks were vehicles with a high profile and took up a lot of space when transported by sea. (TM ORD-9 SNL-G-227)

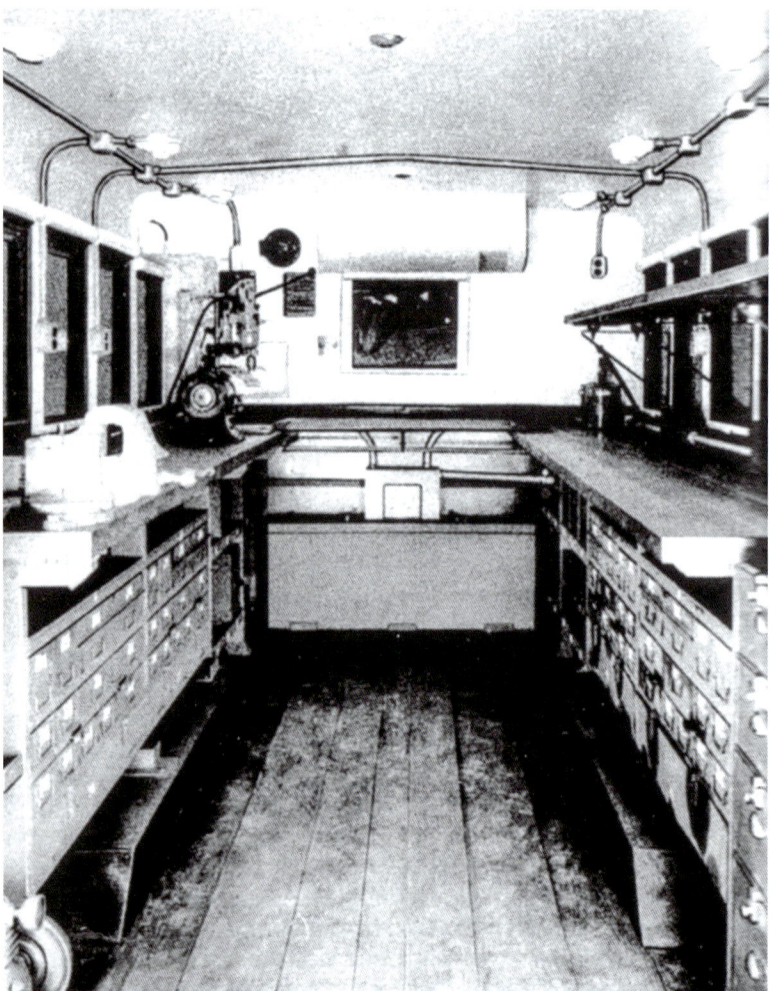

M30 SIGNAL CORPS GENERAL REPAIR TRUCK SPECIFICATIONS

Maker	General Motors Co.
Model	CCKW-353
Powerplant	GMC
Length	300 in
Width	98 in
Height	130.5 in
Min Height	91 in
Wheelbase	165 in
Weight	19,017 lb
Payload	8,007 lb
Gross weight	27,024 lb
Tires	10.00×20 8-ply
Top speed	46 mph
Fuel capacity	45 gal
Range	212 mi
Consumption	4.7 mpg
Brakes	hydraulic

Top: Removing the window strips on the ST-6 body sides meant considerable space could be gained in ships' holds. A simple hoisting system enabling the roof to be lifted and the windows put back in place was enough to return the body to its normal height. (TM ORD-9 SNL-G-227)

Left: The M 30 Signal Corps repair trucks were delivered fully equipped. The apparatus at the top of the partition at the back is the interior heating. (U.S. Army Signal Corps)

By 1942, standardization had become the norm, meaning the steady disappearance of the 1½-ton, 4×4 range of trucks. The jobs of the medium-sized trucks were taken over by the new CCKW 2½-ton, 6×6 trucks of which there were several different technical versions. Among these was the ordnance maintenance truck with a monobloc ST-5 technical body, made by the Superior Coach Corporation. The Army ordered 1,478 examples.

However, this version of the ST-5 body was only a transition in the GMC workshop development. Its imposing body was bulky and took up a large amount of space when transported by ship. This gave rise to the ST-6 body variant, which was unique in that its height could be reduced by removing the side window strip. Some 10,864 of this type of body were mounted on CCKW-353 chassis. Thomas Car Works, Hackney Bros., P&B, and Hicks Body Company delivered the truck/body pairs to the Mobile Shop Depot in Philadelphia. Civilian workers under military management then equipped 1,307 of them as Signal Corps repair trucks.

The workshop body was very specific and was aligned to the tasks that the truck would be carrying out. The internal layout comprised work areas and storage space which each crew organized as to its own needs and tasks. Each truck coming off the production line was fitted with an Evans Products Company heating system that ran on gasoline drawn from the truck's tank. This meant it could operate in all weather conditions.

The inside walls of the body were made of fiberglass panels for complete thermal insulation. Apart from an exterior metallic grille, the windows were fitted with blackout so that the men could work at night. The power for the electrical installation with its various lights and plugs was supplied by a small generator. So as not to betray the vehicle's position, if the doors were opened accidentally, a blackout switch automatically cut the lights.

The right-hand side of the technical body was mainly taken over by radio equipment repairs, the upper shelf taking the instruments and checking and calibrating apparatuses.

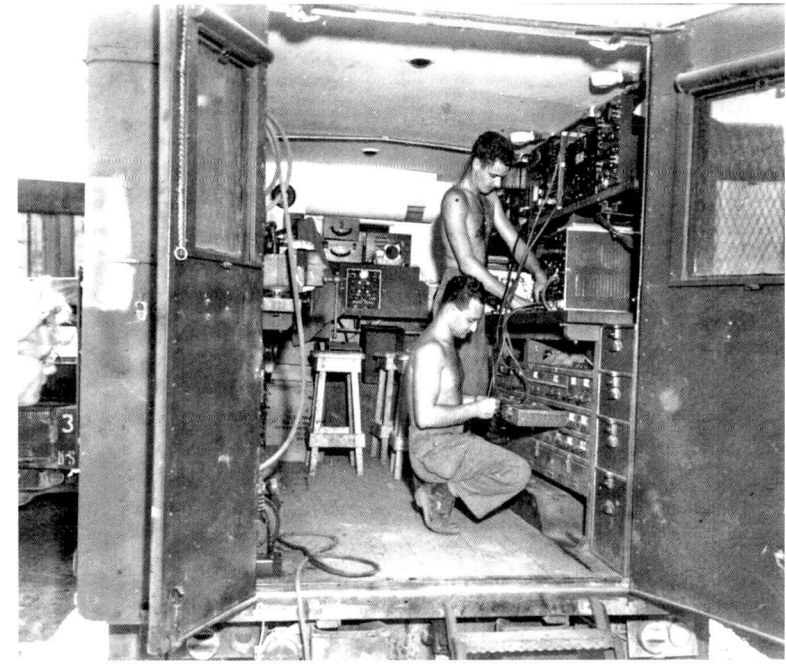

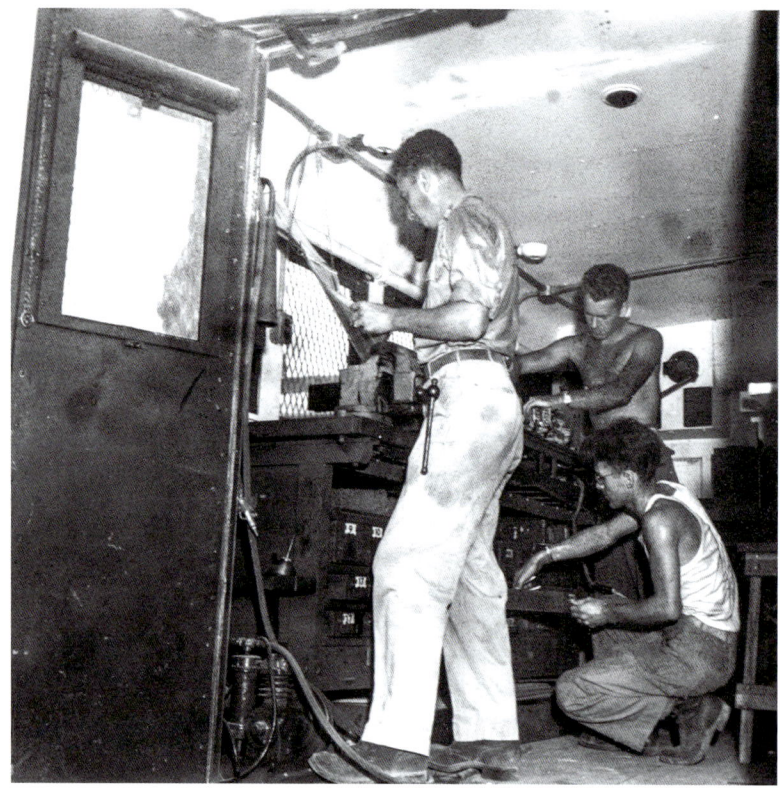

Originally, there were only two small interior lights, turned on and off by a switch located near the rear doors and functioning off the truck battery.

The GMC workshop's availability, its volume, and comfort were such that it was used for tasks it was not designed for. Both during the war and later, it was transformed into, among others, a communications center or even a broadcasting station, all these new uses and developments being more or less elaborate alterations carried out in the field.

Tactical Trucks

Tactical vans carrying a radio were considered as almost anecdotal in the ranks of the U.S. Army. Military commands referenced only 5,645 of them between 1940 and 1943.

The revolution in the domain of the tactical units' radio communications began in 1940. The technological advances financed by the Army were to enable as many of its units as possible to be equipped with direct means of communication.

In the end, all the armored vehicles had to be able to communicate with their different hierarchies. Whether they were scout cars or armored cars, half-tracks or assault tanks, they were all equipped with the elements required for installing a radio. It was important they all produced 12 volts which was the minimum needed for the equipment to work properly.

At the time, infantry still reigned supreme on the battlefield, and it needed modern, mobile communication means. Divisions were therefore equipped with vehicles carrying medium-range transceivers. Initially there

Above: The left-hand side was occupied by a wide workbench equipped with a vise for the larger repairs on the frames and structures of the sets and other signals equipment.

Below: On the island of Guam, repairing the antennae of an SCR-268 radar set, whose K-28 trailer can be seen and on which only the main radar jib still stands. Entirely tooled up for operations of this type, the M-30 was the Signal Corps' main field workshop. (U.S. Army Signal Corps)

were 14 Model AC-101 GMC trucks (USA-W- 206472 to USA-W-206485). The project was viable from a technical point of view but because they were mounted on 4×2 vehicles on a ½-ton chassis, the diminutive AC-101s could not really maneuver on uneven ground, especially if surfaces were slippery.

Reconnaissance and command vehicles were therefore designed, the most famous being the Jeep. As it was more discreet, it finally supplanted the 4×4, ½-ton version of the Dodge trucks.

Five generations of Dodge trucks followed each other with this designation, but none was given a "K" number because they were issued to all service arms and not just the Signal Corps.

Right: Quite used to what all their tasks required of them, those in the maintenance/repair teams were also useful handymen, like here with the fabrication of protective cages. Always present is the indispensable M-30 repair truck. (U.S. Army Signal Corps)

Below: Sometimes it was easier to work outside the truck because of the weight and volume of the equipment needing modification or repair. Here initial tests are being carried out on equipment still installed in the transport cupboards. (U.S. Army Signal Corps)

Some extreme uses meant the furniture inside the body had to be dismantled, like here in Normandy. This M-30 repair truck operated after June 11 in the hinterland and was modified into a long-distance transmission station. It took only 20 minutes between message encryption in France and deciphering in the United States, enabling Americans to follow what was happening on the new operational front in Europe almost in real time. (Plant photo Ref. 30613-30614)

Left: Although it wasn't supposed to be a radio or telephone command post, the M-30 Signal Corps repair truck nonetheless did this work, although there weren't a lot of examples. As they had no antenna base plate or any form of mast, simple wooden perches were used. (U.S. Army Signal Corps)

Below: Whether they were the reconnaissance truck or the radio reconnaissance truck, these Dodges were the only vehicles of this type ordered by the Army. From left: a WC ¾-ton from 1942, a WC ½-ton from 1941, and a VC ½-ton from 1940. Assembled at the Aberdeen Proving Ground, here they have their hoods down. Lowering their outline and widening their track were two of the main features in their evolution. (APG Ref. 362-22-HOMB)

1942
¾ TON 4×4

1941
½ TON 4×4

1940
½ TON 4×4

Model T-202, Dodge ½-ton, 4×4 Truck

Some 2,155 examples of this first-generation truck were produced, of which only 60 were defined as radio reconnaissance trucks under the manufacturer's denomination VC-2; there were three series of registration numbers: USA-W-60430 to USA-W-60434, USA-W-60438 to USA-W-60643, and USA-W-601426 to USA-W-601454.

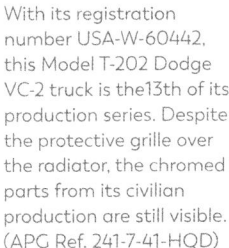

With its registration number USA-W-60442, this Model T-202 Dodge VC-2 truck is the 13th of its production series. Despite the protective grille over the radiator, the chromed parts from its civilian production are still visible. (APG Ref. 241-7-41-HQD)

Fort Oglethrope, January 15, 1941. It wasn't long before the VC-2 was fitted with a 12-volt electrical system. As the large generator took up a lot of space inside the engine compartment, the new battery was placed on the right-hand running board, inside a metal coffer mounted on hinges. (U.S. Army Signal Corps)

They were vehicles derived directly from civilian production, with beautiful, high, harmonious bodywork, made by the Edward G. Budd Company of Philadelphia. They were built on a 4×4 chassis and powered by a 6-cylinder, 3,295-cc engine, rated at 79 bhp at 3,000 rpm. Intended for taking the SCR-193s, the early-production examples were 6-volt, since the radio worked on a battery or via a manual dynamo. This absurdity was soon rectified, the vehicles being fitted with a new 12-volt electric system, installed on the right-hand-side running board.

Model T-207 Dodge ½-ton, 4×4 Truck

The step made between the T-202 range and that of the T-207 was mainly done by simplifying the bodywork and using more robust axles. The new 3,753-cc engine was more powerful and rated at 85 bhp at 3,000 rpm.

Altogether 11,351 examples of this new form of reconnaissance truck were produced. They were all fitted with a 6-volt electric system, except for the 548 examples of the radio reconnaissance truck which were fitted with a 12-volt system to power the SCR-193. These vehicles were defined as Dodge WC-8s and three series of registration numbers were allocated to them: USA-W-60473 to USA-W-60479, USA-W-605100 to USA-W-605304, and USA-W-606021 to USA-W-606356.

Model T-211, Dodge ½-ton, 4×4 Truck

This T-211 model represented a crucial stage in the development of the Dodge ½-ton, 4×4 truck range. There were practically no differences between it and the T-207, except those features of the T-207 that could be improved upon, among which was the braking system that had indeed given rise to a lot of problems during the various 1941 maneuvers.

Top: Although the reconnaissance trucks had a fold-down work surface on the back of the front seat, the radio reconnaissance trucks used it for installing radio equipment. Lieutenant Colonel H. C. Mewshaw of the 1st Cavalry Division checks off his readings on a portable chart, Fort Bliss, July 1941 maneuvers. (U.S. Army S-126449)

Above and left: Registration no. USA-W-606081 applies to this WC-8 Model T-207 radio reconnaissance truck. The impressive 12-volt battery coffer is visible halfway along the running board. Whether they were WC-8s, WC-16s, or Wc-25s, from the outside all these vehicles were alike; only the registration numbers differed. These two shots were taken in August 1941 at Fort McPherson. The Dodge belongs to the 62nd Signal Battalion. (U.S. Army SC-122054 & -122053)

A total of 4,362 new vehicles were ordered for reconnaissance, including 882 radio reconnaissance vehicles. These were WC-16s with the registration numbers USA-W-606357 to USA-W-607238. Six trucks—numbers 607223, 607225, 607227, 607230, 607231, and 607236—were incorporated into the program for the next series, the T-215s, by being fitted with the new Dodge engine for testing.

Model T-215, Dodge ½-ton, 4×4 Truck

Administratively, this series of T-215 Dodges was a transition since the T-214 series was not yet ready. The manufacturer created these vehicles by using the two previous series as a base, fitting a different powerplant and removing all signs of its earlier civilian existence,

Downward view into the cabin where three men can work comfortably. The radio installation is the SCR-193, consisting of a BC-191 transmitter, a BC-312 receiver, and a BD-77 dynamotor. (U.S. Army SC-122055)

DODGE ½-TON RADIO TRUCKS

	Dodge VC-2	Dodge WC-8	Dodge WC-16	Dodge WC-25	Dodge WC-58
Manufacturer	Chrysler Corp.	Chrysler Corp.	Chrysler Corp.	Chrysler Corp.	Chrysler Corp.
Model	T-202	T-207	T-211	T-215	T-214
Orders	60 examples	548 ex.	882 ex.	1,797 ex.	2,344 ex
Powerplant	Dodge	Dodge	Dodge	Dodge	Dodge
Length	188 in	179 in	179 in	179 in	164 in
Width	74 in	76 in	76 in	76 in	78.5 in
Max. height	88 in	83.50 in	83.50 in	83.50 in	81.5 in
Min. height	86 in	68 in	68 in	68 in	62.25 in
Wheelbase	116 in	116 in	116 in	116 in	98 in
Weight	4,277 lb	5,073 lb	5,073 lb	5,117 lb	5,379 lb
Payload	1,000 lb	1,000 lb	1,000 lb	1,000 lb	18,001 lb
Gross weight	5,278 lb	6,074 lb	6,074 lb	6,074 lb	7,183 lb
Tires	7.50×16 6-ply	7.50×16 6-ply	7.50×16 6-ply	7.50×16 6-ply	9.00×16 8-ply
Top speed	54 mph	54 mph	54 mph	54 mph	54 mph
Fuel capacity	15 gal	25 gal	25 gal	25 gal	30 gal
Range	160 mi	300 mi	300 mi	210 mi	240 mi
Consumption	10 mpg	12 mpg	12 mpg	8.4 mpg	8 mpg
Brakes	hydraulic	hydraulic	hydraulic	hydraulic	hydraulic

May 29, 1941. A small ceremony takes place to mark delivery of the 40,000th Dodge to the Army. It is a WC-8 radio reconnaissance truck, like all the others in the front row. (Wide World Photos Ref. M-9438)

like the instruments on the dashboard which was comprised of entirely standard military dials and gauges. A total of 5,646 reconnaissance trucks were delivered under the T-215 designation; 1,797 were WC-25s equipped with a radio registration numbers USA-W-607239 to USA-W-607383, USA-W-607687 to USA-W-607943, USA-W-608358 to USA-W-609007, and USA-W-2071355 to USA-W-2072099.

The registration numbers were exchanged: the radio vehicles' "60" prefix for the "20" of the "Command and Reconnaissance" vehicles. The powerplant was still a 6-cylinder engine but now 3,780 cc, rated at 99 bhp at 3,000 rpm.

Production of the ½-ton radio reconnaissance truck category ended with this fourth model. Technological developments were now biased toward the ¾-ton vehicles (with a 1,650-lb payload). Even though all the previous designs of 1940/1 were specific to a

peacetime America, the next stage of the production process took on a resolutely more military approach.

Model T-214, Dodge ¾-ton, 4×4 Truck

The arrival of the Dodge ¾-ton range was the fruit of a protracted study that drastically lowered the outline of tactical vehicles as the starting point. They appeared in April 1942 and remained operational throughout the war, whereas the ½-ton trucks gradually disappeared. Most never left the United States.

As for the previous series, 37,186 reconnaissance vehicles were ordered, and this time all were 12-volt versions. Only 2,344 were really suited for radio use: the Dodge WC-58s, which were registered from USA-W-20182609 to USA-W-20184952. Compared with the other radio reconnaissance trucks, they

The end of the assembly line with the registration number USA-W-607132 being applied, defining it as a Model T-211 WC-16. A sticker on the windshield confirms it is a radio vehicle.

were entirely new vehicles, broader and lower, with new general axles, and a new 6-cylinder, 3,780-cc powerplant rated at 93 bhp at 3,200 rpm.

To continue the effect of rationalization, the Army only placed one order for these WC-58s, since it did not want further vehicles specifically designated as radio vehicles; it even handed over 650 examples to Great Britain as part of the Lend-Lease.

For all that, the end of the Dodge WC-58 radio trucks did mark the demise of this ¾-ton range as radio

Above: This Model T-215 WC-25 (USA-W-608444) has just been delivered. It still bears its windshield stickers including the one showing it is a radio truck. (APG Ref. 459-1-HQD)

Left: Signal General Development Laboratory, October 13, 1942. The final development of these tactical trucks equipped with a radio was the Dodge T-214 WC-58, ¾-ton truck. Although the vehicle was quite different mechanically, it retained the overall look of the ½-ton series. (U.S. Army SC-147374)

This example is a combination of United States communication means, an SCR-508, and British with a Mk II Wireless Set No. 19. The 508 was a tactical short-range set intended for armored units, made up of a BC-604 transmitter and BC-603. (U.S. Army SC-147375)

A new vehicle did not necessarily mean new radio systems. The SCR-163 remained in general use, even in the WC-58 Dodge. This radio set, derived from an aircraft radio, was adapted specially to work in a vehicle, with a range of 62 miles in Morse code and 15 miles with voice communications, including when on the move. It was only later that the SCR-508 and SCR-608s were preferred for their greater adaptability. (U.S. Army Signal Corps)

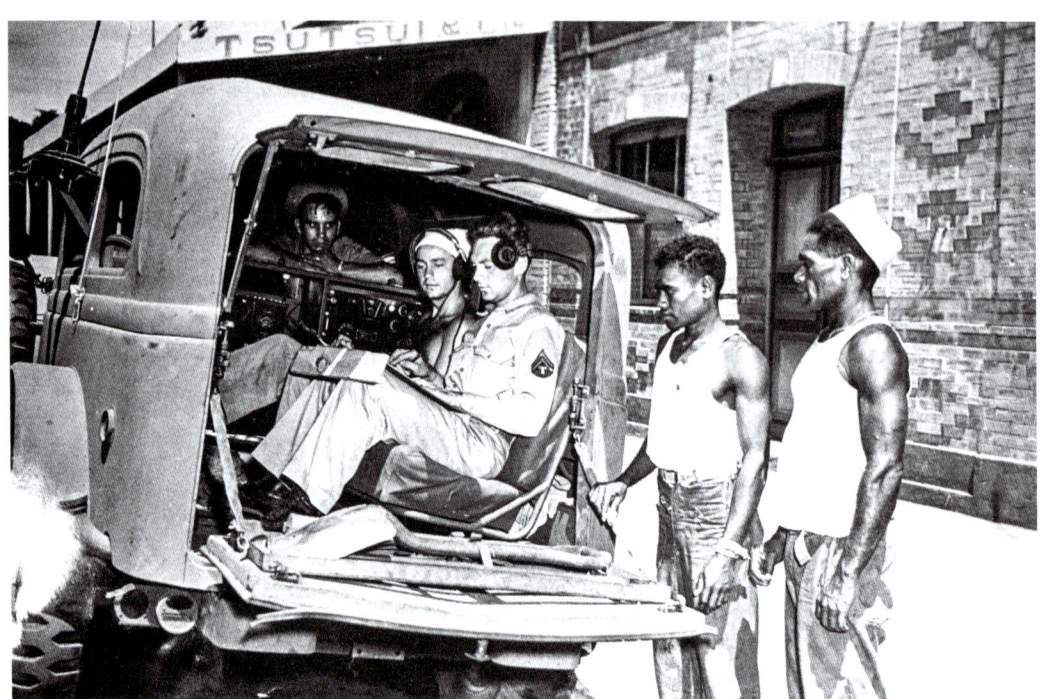

Fort Monmouth, 1943, a typical example of a WC-56 command and reconnaissance truck, as the "20" registration prefix on the hood tells us. It is used as a radio reconnaissance truck. The worktable has not been dismantled and the radio operator appears cramped. The SCR-608 was an artillery set comprising a BC-684 transmitter, two BC-683 receivers with 10 preselected channels, and a range of 10 to 15 miles for voice communication. (U.S. Army Signal Corps)

Another secondary use for this Model T-214 4×4, ¾-ton Dodge WC-53 Carryall: designed as a multi-purpose vehicle, its interior was adaptable and could therefore take this installation. From the outset, the vehicle was mounted with a 12-volt system and a base plate for the antenna has been added on its left-hand flank. Altogether 8,400 examples were manufactured over a very short period from late 1942 to early 1943. It was confined to the rear of the front lines. (U.S. Army Signal Corps)

trucks. The 34,842 WC-56s (and WC-57 for those equipped with a winch at the front) tactical command vehicles for officers (command and reconnaissance trucks) could indeed be fitted with a transceiver with a bit of transformation. This also happened to another truck: the WC-53 Carryall, also ordered in a 12-volt form so it could, if needs be, be fitted with a radio.

Radio Jeeps

The unofficial designation Light Radio Reconnaissance Truck was nothing else but the small 4×4, ¼-ton general-purpose (GP) vehicle built by Bantam, Willys, and Ford and better known as the "Jeep."

The Jeep's basic function was not as a radio vehicle; it was not equipped for that particular use: the electrical system was basic, and comprised elements working with or producing 6 volts. It went on, however, to become a vital link in the Army's wireless network even if the Signal Corps did not allocate it a Model K number.

Its debut in this role was no doubt the conjunction of a series of factors which, if taken separately were inconclusive. It was designed early on, it was small, and it was widely distributed; these factors meant that it had advantages which the radio reconnaissance trucks described above did not have.

Very quickly modifications appeared in the field as the DIY system was in full swing. The Jeep carried sets which worked only with batteries, or with 6 volts, or which could be powered by a manual generator, like the SCR-284. Some units even installed small generators on the front passenger seat.

How the Radio Was Powered

In its Fort Monmouth laboratories, the Signal Corps tried to transform the Jeep by installing a 12-volt system and adding a voltmeter with its switch on the dashboard, like the ½- and ¾-ton trucks. This modification was not accepted, neither was it included on some of the Jeeps yet to be produced, probably due to the costs not being justified.

in which a terminal box was installed, or to anywhere else depending on the installer's imagination and the place chosen for the radio.

- The second kit was bigger, and its aim was to transform the 6-volt electrical installation into 12 volts. It comprised a 12-volt generator, a voltage regulator, and a second 6-volt battery to be installed in parallel, and everything needed for positioning the terminal box.

- The third option was more sophisticated and was carried out with a power take-off (PTO); it was intended for the bigger radios like the SCR-193. In fact, it required a generator that was far too big to be installed inside the engine compartment. Placed between the front seats, this was driven by pulleys and belts taken off the rear prop shaft up through the floor.

A Standard Connection

All these modifications, especially installing the terminal box and the electrical cable, took time.

Around March 1943, Willys and then in May, Ford, installed a radio outlet box as standard on the right-hand side of the body. This set-up greatly facilitated adapting the Jeep as a radio vehicle. There were many adaptations, often Jeeps with large radio installations and often with sets used by armored

There were, however, four solutions for the users: they could continue to carry out makeshift transformations to adapt their Jeeps or use one of the three available conversion kits:

- For basic uses, the first kit solution comprised simply an electric cable taken off the 6-volt battery to the chest behind the passenger seat

Morocco, March 13, 1943. Here the radio has been installed in place of the back seat of this light vehicle. (U.S. Army SC-380224)

vehicles. Expedience therefore dictated that vehicles in this layout became radio vehicles rather than reconnaissance vehicles.

On the Dodge Trucks

Various needs in the field also meant that the Dodge WC-51 and WC-52 weapons carriers were transformed into radio vehicles. It was easier to carry out the 12-volt changes since there was more space and the 12-volt components—like the generator and the batteries—were already part of the WC-53s, -56s, -57s, and 5-8s which were factory mounted with this voltage. Besides, as demand was high, during a factory reconditioning program at the end of 1944/early 1945, almost 4,000 of these ¾-ton Dodges were fitted with a complete 12-volt system.

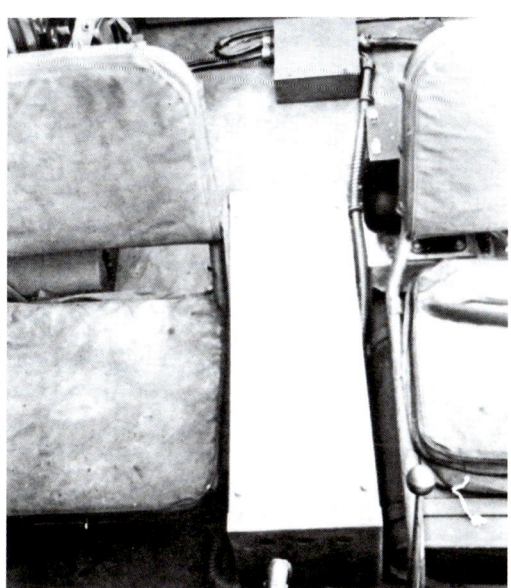

Above and right: Making a Jeep into a radio vehicle like the ½-ton or ¾-ton trucks left almost no space. There was just no room for any additional equipment, and even less for the crew's personal effects. Only a trailer could make up for this lack of space, but hitched up like this, the vehicle lost all the advantages of its diminutive size and its great mobility. (U.S. Army SC-146182/146183)

Above: The BC-312 radio receiver, the TU-18 tuning unit, and the LS-3 loudspeaker have been installed over the right-hand wheel arch. An MP-37 antenna base plate has been set up on the left-hand side of the bodywork, protected by a sloping anti-branch protection. The large battery chest is between the two front seats. (U.S. Army SC-146184)

Above: The primary element, the BC-191 radio transmitter, has been installed on the left-hand wheel arch; in front of it is the PE-55 dynamotor generator. This layout left barely enough space for the operator who was wedged in the back seat without any real possibility of getting out quickly. (U.S. Army SC-146185)

England, March 1944, the 286th Joint Assault Signal Company (JASCO) has set this Jeep up with a double radio. An SCR-284 on the right looked after communications with the infantry whereas the SCR-610, on the left, was an artillery post. (U.S. Army SC-293660)

Left: Another view of the 286th JASCO Jeep. JASCOs were attached to the Engineers' special brigades deployed during the landings. They ensured inter-unit communications as well as liaison with naval and air support. (U.S. Army SC-293662)

Above and left: The SCR-284, although it could work using different electricity sources, was accompanied by a number of accessories and occupied practically the whole of the right-hand half of the back of the Jeep, leaving precious little space for the transmitter. (U.S. Army SC-293664)

Above left: The very adaptable SCR-284 was widely used by the infantry and field artillery. As it could be powered by batteries, by a PE-103 generator, by a manual generator (at left in the photo), or by a vehicle by going through a 6- or 12-volt PE-104 converter, it could be transported in a variety of ways. The equipment on the right is a remote control. (U.S. Army SC-257531)

Above right: Camp Gordon, February 23, 1943. The complete SCR-284 weighed 110 pounds which justified it being mounted in a Jeep rather than being manhandled. It was placed on the floor and did not require complicated equipment to secure it: simply folding down the back seat did the trick. However, the vehicle could only carry two crew members. (U.S. Army SC-257531)

Left: Batignolles Station, Paris, February 15, 1945. Private First Class Bright Sink, back from a surveillance patrol of the railway installations, reports to two military policeman: T/4 Louis C. Pelsor is the radio operator and T/5 George A. Morris the driver. (U.S. Army SC-200845)

Above: On this installation, the SCR-284 is mounted in an offset position at the back of the Jeep, which has freed up the rear bench seat. The roof tarpaulin has been modified (back right) to protect the equipment. (U.S. Army Signal Corps)

Left: The power supply for the series of SCR-610 artillery sets and other similar models was flexible: batteries, generator, or vehicle battery. Its small size enabled it to be mounted over the wheel arches, but its 5-mile range was a limitation. (U.S. Army SC-257532)

Left and below:
In-unit modifications were sometimes quite sophisticated as shown with this Jeep: double radio sets, trailer with generator, and makeshift roof tarpaulin with extra canvas flaps for a complete blackout. The markings of the 511th Signal Company, an organic signals unit of the 11th Airborne Division, can be seen on the vehicle. (U.S. Army SC-241052/3)

7

Specialized Trailers

Among the "K" models in the Signal Corps was a series of trailers for specific uses. From the ¼-ton to 5-ton models, all mounted on a single axle, they provided technical support in their specialized fields.

K-36 Telephone Construction and Pole-Hauling Trailer

February 1945, Livorno, Italy. Even though it was designated to be towed by the 1½-ton Chevrolet or a GMC CCKW-2½-ton truck, the K-36 could be towed by almost any other truck. (U.S. Army SC-202908)

The K-36 telephone construction and pole-hauling trailer, or in military parlance, Trailer, Telephone Construction and Pole Hauling, K-36, was a small transport platform trailer with an extendable drawbar. Built by American Coach & Body Company and Highway Trailer Company, among others, its main function was to transport lengthy items like telegraph poles. Comprising a chassis frame with a drawbar and a fixed axle on which were mounted 7.50×20 8-ply tires, it had a flat bottom for normal transportation. Its accessories included four small removable slatted sides, two of which had mudguards. It was fitted with four foldable uprights which once in position could slide out to hold the goods being transported in place. This trailer was not equipped with any braking system apart from a parking brake.

K-37 Telephone Construction and Cable-Hauler Trailer

The K-37 trailer was specially designed for reeling out large coils of telephone cable and transporting long items. Built by Highways Trailer Company, among others, it was made up of a chassis frame with a drawbar and a fixed axle on which were mounted 9.00×20 10-ply tires. It did not have a flat bottom like the K-36 trailer. Two forward foldable side uprights could slide out along an axis to hold longer items in at the side. A second set of these side posts could also be fitted. Two posts with a crossbar served as a reel for the cables. Fitted with electric brakes, it did not have an extendable drawbar.

K-38, Telephone Cable-Splicer Trailer

Built by Highway Trailer Company, York-Hoover Body Company, and FWD (Four Wheel Drive) Auto Company, among others, the K-38 trailer was specially designed to take the equipment needed for splicing

Top right: England, September 30, 1942, this K-36 telephone construction and pole-hauling trailer is shown folded up. The drawbar is retracted, the small side rails are not installed, and neither are retention arms. (U.S. Army SC-136174)

Right: Signal Corps laboratories, Fort Monmouth, June 12, 1942. With its drawbar extended, this K-36 trailer has all options installed, including the four small wooden side rails. (Signal Corps photo SCL-4279)

Below: Tinian, Mariana archipelago in the Pacific, the 443rd Signal Heavy Construction Battalion is setting up equipment. A K-43 Chevrolet tows a K-36 trailer loaded with telegraph poles. (U.S. Army SC-380111)

SIGNAL CORPS SPECIALIST TRAILERS

	K-36	K-37	K-38	K-45	K-52	K-55	K-63
Category	2-ton	5-ton	¼-ton	1½-ton	1-ton	1½-ton	1-ton
Length	121.5 in[3]	146 in	78 in	198 in	140 in	295 in	140 in
Width	72 in	90 in	38.5 in	83 in	67.5 in	95 in	67.5 in
Height	36 in[4]	68.75 in	42 in	101 in	112.5 in[1]	115 in	731
Weight	1,816 lb	2,901 lb	410 lb	4,755 lb	1,283 lb[2]	8,208 lb	1,283 lb[2]
Payload	484 lb	10,003 lb	500 lb	3,000 lb	2,000 lb	4,503 lb	2,000 lb
Gross weight	6,821 lb	12,910 lb	910 lb	7,7528 lb	3,285 lb	12,712 lb	3,285 lb
Tires	7.50×20 8-ply	9.00×20 10-ply	4.00×18 4-ply	7.00×20 6-ply	7.50×20 8-ply	7.50×20 8-ply	7.50×20 8-ply
Brakes	w/o electrics	w/o	w/o	w/o	w/o	w/o	w/o

[1] Total height with the tarpaulin on the supports.

[2] Weight of the trailer with the metal body, the wooden one being slightly heavier.

[3] Length with the extendable drawbar retracted.

[4] Height with both side rails and uprights installed.

A signals construction company in action. No fewer than six men were needed for handling a telegraph pole. The K-36 telephone construction and pole-hauling trailer was a vital piece of equipment for line laying. (U.S. Army SC-175387)

The K-37 telephone construction and cable-hauler trailer was large and robust. Although it was designed mainly for carrying large reels of telephone cables, it could also be used for transporting long objects thanks to its four foldable upright arms. (U.S. Army SC-145270)

Highway 54, Manila, Philippines. Laying a new telephone line along the road that encircled the capital. This K-37 trailer is towed by a K-43 Chevrolet truck. (U.S. Army SC-255425)

In March 1942, the Army was preparing to connect Puerto Rico to the rest of the world with a sub-marine telephone multi-cable. The Signal Corps was entrusted with the land element of the task. These heavy Western Electric reels were handled using K-37 trailers. This easy-to-use trailer did not need any lifting apparatus for putting the reels in place. Getting them back horizontally was done by the towing vehicle's winch, the towing cable of which is visible here. (U.S. Army SC-131877)

Right: Land operations for laying cables under way; Sergeant Harold D. Reed has opened the reel. Only the weapon in his belt and the K-37 trailer indicate that he is not a civilian. (U.S. Army SC-131872)

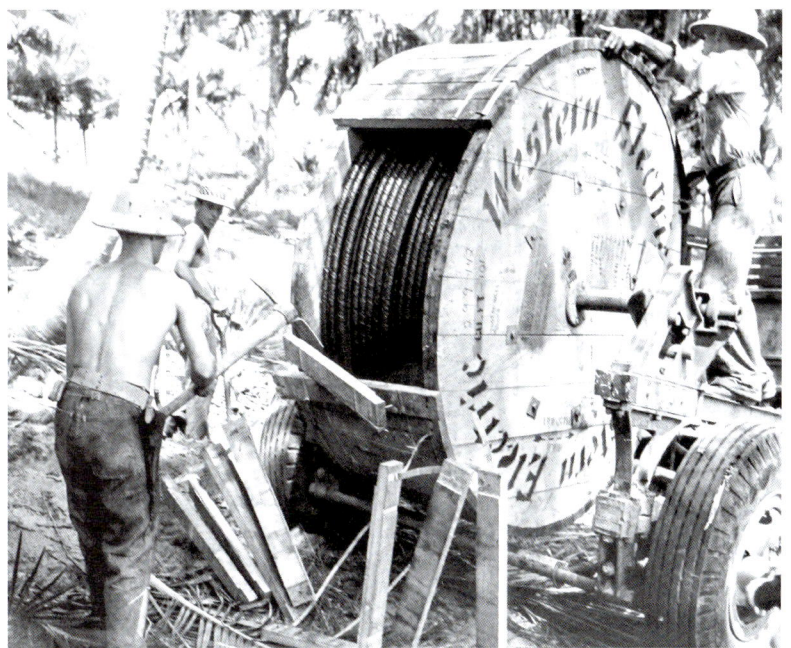

telephone cables for both new installations and repairs. It consisted of a chassis frame with a drawbar, a fixed axle on which were mounted 4.00×18 4-ply tires, and an elevated compartmentalized trunk.

The main part was accessible from above by lifting the cover panel; it contained all the tooling. The chest fitted between the leaf springs had a small access panel at the back to reach the bottom of the trailer and a small stock of cable.

There were several different details depending on the manufacturer: a lower chest or not, a rear fold-down table, a protective cover for the road lamp on the top, and so on.

K-45 Photographic Van Trailer

These photographic trailers are anecdotal because only a few dozen examples existed. They came directly from the civilian production of the Fleetwood Gilder Trailer Company. Their purpose was enabling development of photographs taken by Signal Photo Company photographers.

They were very quickly declassified because they were fragile and not suitable on campaign. They are today wrongly assumed to have belonged to the Corps of Engineers, including Map Reproduction Units.

These laboratory trailers were built on a chassis frame with a drawbar and a fixed axle on which were mounted 7.00×20 6-ply tires. The interior was fitted with work surfaces, drawers, a water supply for an electric water heater and shutters for closing off all the windows.

K-52 Power Engine Trailer

The Signal Corps, which used a lot of electricity on all its equipment, also used normal trailers for setting up its generators.

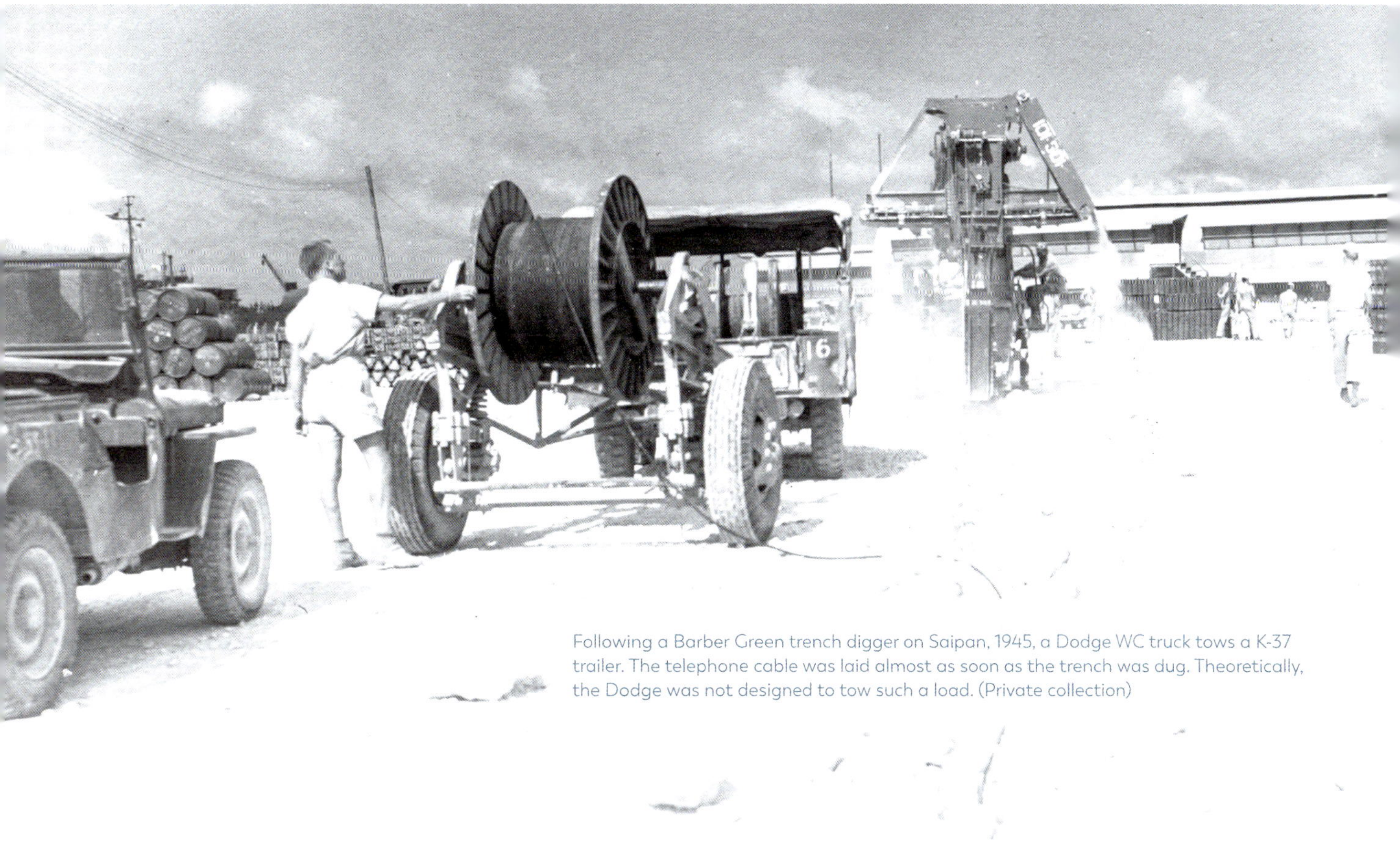

Following a Barber Green trench digger on Saipan, 1945, a Dodge WC truck tows a K-37 trailer. The telephone cable was laid almost as soon as the trench was dug. Theoretically, the Dodge was not designed to tow such a load. (Private collection)

When they were attached to AA defense units and were not accompanied by specific trucks for producing any current, the AAA M7 or M8 trailers were used for this purpose.

Three groups of generators—power engines, or PEs—were involved. Considered as the second part of the pair with the K-51 truck, the K-52 generator PE-95 trailer was used more often when a standard-capacity generator was needed. The 1-ton, 2-wheel trailer was used to carry it.

259,064 examples were delivered to the U.S. Army, of which a very small number appeared in the Signal Corps nomenclature of K-52 trailers.

The PE-95 came in three large families:

- The Types A, B, and C mainly comprised a Ford engine rated at 20 bhp at 1,200 rpm, and a generator made by D.W. Onan & Sons. It

Right: The removable wire-mesh cage (lantern guard) which protected a sidelight, lit when on the move. (TM SNL G-685)

Left: Built by FWD Company and Highway Trailer Company, the K-38 telephone cable-splicer trailer was most certainly the smallest trailer on the list of "K" models. (U.S. Army SC-145269)

Above: Being very narrow, the K-38 had a poor reputation. Normally towed by a K-50 truck, it could topple over if the driving got too vigorous. A Jeep would have been more suitable for its ¼-ton size. (TM SNL G-685)

Right: The K-38 trailer was entirely compartmentalized and looked like a big toolbox. Each section housed the parts needed for making splices and watertight connections. (TM SNL G-685)

Above: England, 1944, the small K-38 trailer. Driver agility was needed when it was hitched up behind a truck. (Private collection)

was rated at 5 kW for a 115-volt current and 54.8 amps. These types were generally associated with, among others, the SCR-292, -299, -399, and -499 very long-range sets, and the very powerful -696 and -698 sets for propaganda broadcasts used by the Psychological Warfare Service.

- The PE-95 type Fs were based on a Willys engine rated 24 bhp at 1,200 rpm with a generator from D.W. Onan & Sons. Power was 5 kW for a 120- or 240-volt current and 52/26 amps. This variant was replaced by the PE-56-G.
- The PE-95 types G and H were mainly made up of a Willys engine rated at 35 bhp at 1,800 rpm with a D.W. Onan & Sons generator. Power was 10 kW for 120- or 240-volt current and 104/52 amps. They were used just like the A, B, and C models.

These trailers had no spare wheels since they were the same as those on the Chevrolets and GMC CCKWs, the preferred tractor vehicles.

Above left and left: The photographic laboratory trailer, although excellent for developing photographs, was a concept that was quickly abandoned as it proved impractical in the field. (U.S. Army SC-118616/17)

Below left: Generator-carrying trailers in the Signal Corps were manufactured with the 1-ton cargo trailer as a base. Some 259,064 examples were made by 26 different companies. (U.S. Army SC -127237)

Below: There were several arrangements of the K-52, each defined by a letter from A to E. Here it's a type E. Only minor details, their usage, and the towing vehicle differentiated them. (TM- 11 -281)

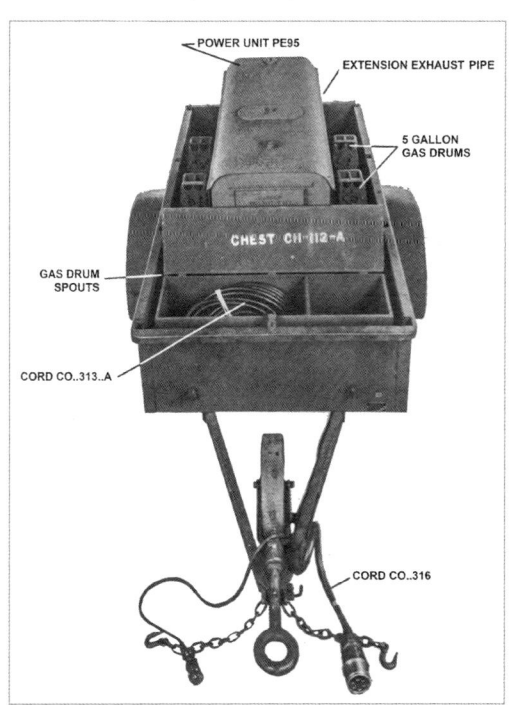

A K-52 trailer with a PE-95 generator set up as a fixed post, Saipan, 1944. (U.S. Army SC -319567)

Refueling a PE-95 generator in the 77th Division. The fuel tank took 10½ gallons and, depending on the model, had a consumption rate of between 1 and 1½ gallons an hour. (U.S. Army SC -169395)

Below: Somewhere in New Guinea is a K-51/K-52 pairing. The ventilation for the generator was provided by opening the rear and front sides of the trailer's tarpaulin. The exhaust pipe passed along the floor under the PE-95 and was extended well out to the rear to expel the toxic fumes. (U.S. Army SC-182759)

The truck's spare wheel was also used for the 1-ton trailer. However, when used with the K-51 truck (see Chapter 4), the truck had no space for this fifth wheel, so it was lodged in the trailer. In many cases, the crews themselves made up spare wheel mountings under the trailer.

K-55 Van Semi-Trailer

The large K-55 was considered as a general-purpose trailer, able to house various items of radio equipment, or as a meeting room or mobile command post.

Built by A.J. Miller and Oneonta Linn, the K-55 trailers were vans on a chassis borne by two axles with single 7.50×20 8-ply tires. They were entirely enclosed and had two large rear doors, a lateral access door, and windows with shutters. There were various versions, classified from B to E throughout production. They were, uniquely, fitted with an electric braking system, in keeping with the K-73 Chevrolet 4×4, 1½-ton tractor truck.

K-63 Power Engine Trailer

This 2-wheel, 1-ton trailer was used for transporting other generators. Fairly uncommon, these PEs did not have specific materiel like the K-52 trailer; they were all put together in a trailer adopting the designation K-63 despite its varied uses. It housed mainly:

- PE-92 types A and B consisting mainly of a Waukesha engine, rated at 19.5 bhp at 1,200 rpm and a Marble Card generator. Its power was 6 kW for a 110-volt current and 68 amps. These PE-92s were intended for use with mobile radio repair shops.
- PE-99 with the designations A to G, all mounted with a Wisconsin engine rated at 20.5 bhp at 2,200 rpm, their power was 7.5 kW for a 120-volt current and 36 amps. The various models were distinguishable by the use of different generators.
- The PE-197Cs were mainly made up of a Hercules engine rated at 13.5 bhp at 1,800 rpm and a Hobart Brothers generator. Power was 5 kW for a 120-volt current and 52.5 amps. These PR-197s were generally used with, among others, the SCR-572, -573, -575, -642, and -643 (air force radios for controlling fighters on the ground).

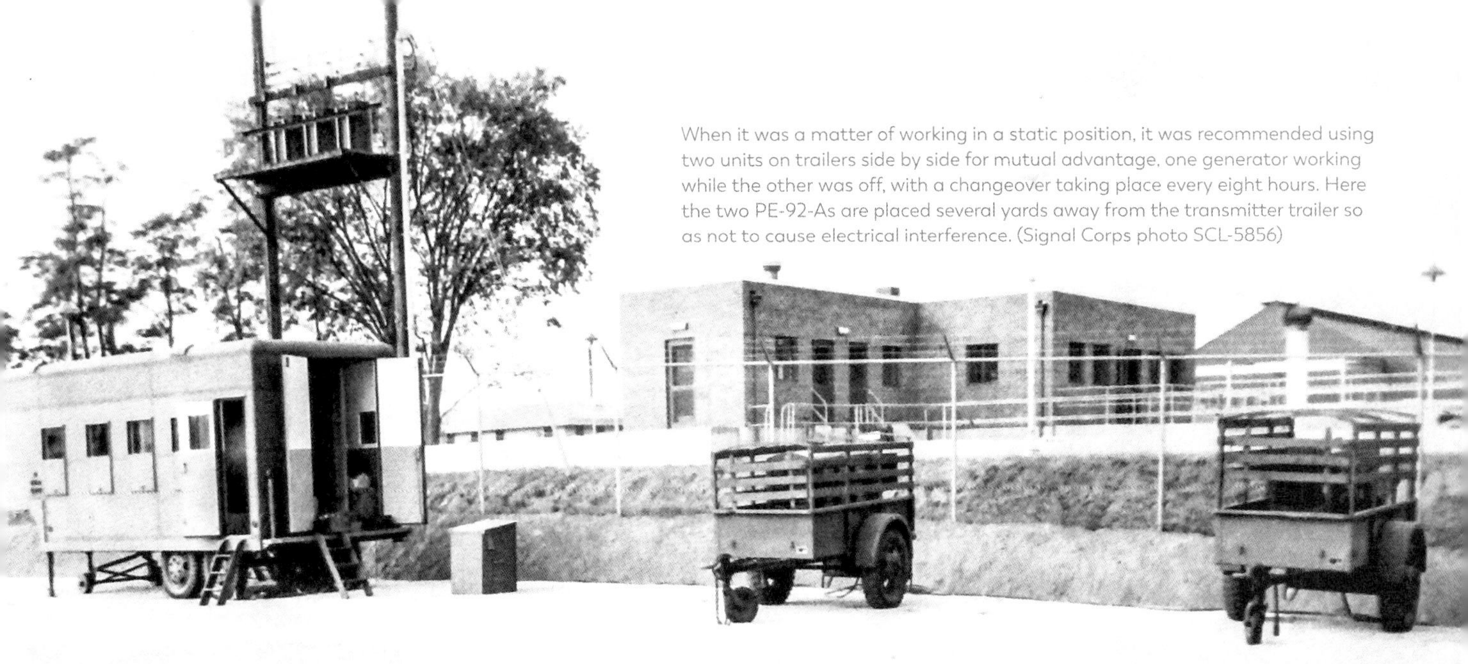

When it was a matter of working in a static position, it was recommended using two units on trailers side by side for mutual advantage, one generator working while the other was off, with a changeover taking place every eight hours. Here the two PE-92-As are placed several yards away from the transmitter trailer so as not to cause electrical interference. (Signal Corps photo SCL-5856)

Left: The K-63 version of the 1-ton trailer was only used for carrying generators without specific materiel. Here it is a PE-92. (U.S. Army SC-142210)

Below: The same set-up with two K-63 trailers carrying PE-99s, in the middle of the Burmese jungle, June 28, 1944. For improved access and ventilation, tarpaulins have been put up to form a sort of tent over the two trailers. (U.S. Army SC-274281)

8 Telephony

Manual Reel Units

Below: The RL-31 reel unit was the primary manual winder/unwinder for the big drums of DR-4 and DR-5 telephone cables; it weighed more than 200 pounds. (U.S. Army SC-120537)

Below right: To help with handling the reel unit, different techniques were described in the manual, but not always too helpful with mobility. (TM 11-362)

Despite lightning advances in wireless transmission, the telephone remained essential for communications in the U.S. Army. Three major methods of laying the lines were used in the field.

The simplest (and toughest) way was to pull the telephone cables by hand. Cables were delivered on large spools and the soldiers faced the task of handling them by hand, no matter what the work conditions were.

Apart from defining the different types of cables, the Signal Corps supplied various manual rolling/unrolling machines called reel units.

Apart from the small RL-39 model used by front-line units, the most common was the RL-31 which used DR-4 spools that carried over 2,600 feet of cable, and the DR-5 with twice that length. The DR-5 was a sort of trestle, placed on the ground. It was heavy and cumbersome and weighed 46 pounds empty and 200 pounds with 5,280 feet of cable. According to the TM 11-362 technical manual, all you needed to do was push the reel like a wheelbarrow to unroll the cable, and although this method was possible on flat ground, on rough ground it became well-nigh impossible. To change into off-road mode, the trestle could be opened to 180° and carried by two men like a stretcher, a solution which very quickly revealed its limitations.

When installing radios on the Jeeps, Signal Corps troops showed how imaginative they could be with the little 4×4 vehicle. Installation kits were created for adapting modifications in the field and instructions were given for these DIY changes. It was also common to affix the RL-31 to the deck of bigger vehicles, like the Dodges and GMCs.

Mobile Reel Units

We have seen how cables were laid by the users at the end of the line. The higher echelons also used reel units, but these were models with an independent motor for heavier work.

As with the manual models, the powered reel units were available in various versions. The most common was the RL-26 and its various developments. Although designed to be used on the ground, it very soon had to be mounted on a vehicle to speed up line-laying. It weighed 435 pounds without the two DR-5 spools,

Wheeled variants which could be either pulled manually, or towed behind a small vehicle, were also available but were commonly thought to be difficult to handle, like this pre-war RL-16 reel cart, or the similar RL-35 reel cart used throughout the conflict. (Credit Line Ref. 568279)

Above and right: Instructions for installing the RL-31 on Jeeps were not followed too closely, the signallers often preferring to improvise, like here, at Camp Chaffee in 1942. This Jeep from the 69th Field Artillery Battalion is carrying a reel unit partly wedged between the body and partly on the floor. (U.S. Army SC-169431 & -169434)

Above: August 28, 1944, near the village of Ponsacco in Tuscany, near Pisa, elements of the 370th Infantry Regiment unroll a temporary telephone line directly into a ditch. The reel unit carrying a DR-4 drum was simply installed on the back floor of the Jeep. (U.S. Army SC -194733)

Above: WC 51/52 Dodges were also much in demand by Signal Corps units because they were small and well suited to their jobs. The reel unit is the RL-31, recognizable by its moulded parts joining the legs to the horizontal reinforcements. It carries two DR-4 spools. (U.S. Army Signal Corps)

Above: Ipil, Leyte Island, Philippines, December 8, 1944. With chains on the wheels, this overloaded Jeep is carrying a ready-to-use RL-31 reel unit amid all the GIs' personal effects. (U.S. Army SC-198043)

Rgiht: Working from the freight deck of a GMC CCKW-353 truck on a road in Shadazup, Burma, June 28, 1944. Private First Class McHenry Petruisch and Private First Class William Kyzbert from the 96th Signal Battalion each handle a reel unit, the former an RL-17 and the latter an RL-31B-C recognizable by the triangular pieces of metal strengthening the joints, the support legs, and the horizontal bars. (U.S. Army SC-274282)

Below: January 27, 1945, near Hagenau, Alsace, France. Private First Class Thomas E. Mais of the 101st Airborne Division lays a temporary telephone line. An RL-31 has simply been placed on the last strut of the chassis, on the rear of the Jeep, and is held in place by two leather straps. (U.S. Army SC-199553)

Left: Belgium, January 8, 1945. Private B. L. Novak, T/5 C. Manos, and T/5 T. D. Jeffers of the 84th Signal Company have just laid a telephone line from an RL-31 to an artillery forward observation post. (U.S. Army SC-198766)

Below: Early "K" models remained in use for troop training, like these horse-drawn reel carts at Camp Fuston in October 1942. They can be considered as the forerunners of the RL-26 reel units. (U.S. Army SC-138202)

Placed on the ground, the RL-26 is of not much use because its 435 pounds prevented it from being manhandled. Here it is being used to put DR-5 drums back in order, rewinding them correctly for future use. (U.S. Army SC-122388)

Laying temporary telephone lines was a simple matter: a vehicle carried the reel followed by the men on foot or in vehicles who hooked the cable up in the trees, or whatever supports were available. When this was not possible, the cable was simply left on the ground. (U.S. Army SC-148712)

so it had to be transported. Made up of an aluminum chassis for the RL-26, and of steel for the RL-26-B, this type of reel unit was fitted with a Lauson or Briggs & Stratton 147-cc, single-cylinder engine rated at 1.75 bhp at 3,200 rpm.

Gasoline was fed to it from a 2½-gallon tank. Controlled from the start by two slip clutches, the reel unit laid the cable at a rate of 30 mph and rewound it at 4 mph. One spool, or two simultaneously depending on requirements, was used for the job.

Some units like the 2nd Armored Signal Battalion (seen here) created vehicles specifically for this task. These arrangements enabled the system to unroll the cable rearward by the forward movement of the vehicle, but also, as here, winding in the cable from the front with the forward movement of the vehicle. (U.S. Army SC-166327)

No fewer than 24 DR-5 spools are being transported by this GMC CCKW-353 truck—some 12 miles of double lines, or 24 miles of single line. (U.S. Army SC-166328)

Above: The 95th Signal Company, 95th Infantry Division, March 27, 1942. As happened quite often, there were no limits to imagination, and anything was done to make things easier. Here the tailgate of the cargo body has been replaced by a hinged running board. (U.S. Army SC-175813)

Right: While Privates Fred Hanunfelter and Anthony Cosemanto handle the RL-26 reel unit, Staff Sergeant Tony Minozeski is busy installing a line. All three belong to the 56th Signal Battalion. Speed was the name of the game for this type of operation; the men were not there to be finicky and attached the lines as best and as quickly as they could. (U.S. Army SC-169224)

North Africa, February 1943. Signaler Captain G. S. Wuterman next to a line-laying vehicle. Note the upright cradle for a light machine gun on the back left-hand side of the body. (U.S. Army SC-171330)

The Signal Corps units that put these RL-26s into operation used the vehicles their units had been issued with, so Dodges, Chevrolets, GMCs, and the like. Very often the equipment was simply just bolted to the floor of the cargo body, and sometimes in pairs with another RL model. In some cases, these adaptations reflected the will to create an almost "professional" vehicle, such was the sophistication and the complexity of the layout of the cargo body.

As with the 12-volt radio Jeep, the Signal Corps tried to create a specific Jeep for laying telephone lines. To do this it used a modified RL-44 reel unit, set in motion by a chain linked to the rear transmission of the little 4×4, through its floor.

A seat for the operator was installed over the back left-hand wheel arch; a wide running board covered the rear and a rack for storing a DR-4 spool was located between the two seats of the cabin. The spare

Below: "K" model or not, it is impossible to tell. This Jeep was nonetheless a creation of the Signal Corps General Development Laboratory, the photo dated September 22, 1942. The RL-44 reel unit is installed on the floor of the rear section. The drive is transmitted mechanically, and a seat has been installed specifically to handle the equipment. There is a wide running board at the back, which means the spare tire is on the hood. (U.S. Army SC-142224 & -142228)

wheel was fastened on the engine hood. Several jeeps were modified in this way, but this initiative was on a relatively small scale.

The Mobile Telephone Switchboards

Curiously, nothing was ever organized for transporting the large telephone communication switchboards. Here more than anywhere else, "DIY/Made in the USA" reigned supreme.

With dimensions varying with the number of lines they dealt with, the telephone exchanges were of capital importance for the war telephony networks. Although the smaller models for tactical units were so basic,

Above left: From the simplest to the more complicated, from the temporary to the permanent, producing these mobile telephone switchboards meant the vehicles became very specialized. On the side of this Dodge VF truck, is a complex connection table mounted by the 62nd Signal Battalion, 1941. (U.S. Army SC-122052)

Above: Linnich, Germany It is difficult today to imagine that victory in 1945 was in large part due to the 250,000 miles of telephone lines laid by GIs, especially when one appreciates how fragile the temporary lines were. Here T/5 Michael Kutcha is carrying out repairs in the middle of a veritable spider's web. (U.S. Army SC-201461)

Left: These vehicles were defined as mobile switchboard units when prepared in-unit and, when there was no urgency, some were almost luxurious. (U.S. Army SC-122051)

they could be transported on the back of a truck, the larger exchanges with several hundred lines were not designed for being moved around.

The signals units which installed them often improvised so they could work in the best conditions. Some somewhat extravagant designs saw the light of day, and trucks and trailers, small and large, became makeshift telephone exchanges. There were no limits to imagination in this field where DIY was the norm.

Above: Tambourine, Australia, September 21, 1942. Sergeants Karl Rhinehard and Marvin Williams and Corporal Elton Hamal operate a telephone switchboard of the 32nd Infantry Division. (U.S. Army SC-168663)

Below: North Africa, 1943. The essential element of the telephone network was the switchboard, no matter the size. (U.S. Army Signal Corps)

Top: July 1944, Shadazup in Burma, a Signal Corps unit has set up and camouflaged its telephone switchboard aboard a GMC CCKW-353 truck. Everything is homemade including the chairs. (U.S. Army SC-274289)

Above: The 56th Signal Battalion GMC 6×6, 2½-ton truck was usually chosen for telephone switchboard usage as its cargo body's large surface made it simple to install. Here the tarpaulin support bars have been raised so the men can stand up. (U.S. Army SC-130662)

Installing Landlines

Another important element of telephone networks, landlines developed in the rear needed to access the front lines for command-and-control purposes.

Being able to install fixed land telephone lines on telegraph poles was a priority which only appeared later in the Signal Corps' list of duties. It was only in 1939 and on a large scale that the Corps took under its wing what had hitherto been entrusted to civilian companies in the United States, even in the case of military installations. The first signals construction battalions appeared and directly copied the work methods of civilian companies, including the equipment they used. As a result, several specialist vehicles plugged the gaps in the Signal Corps "K" list just prior to the war.

Left: Almost everything is already present on these early models, from the work position and the various accessories, to even the right-hand-side exit point for the winch located behind the cabin. (U.S. Army SC-11054)

Below: On the initial listed series of the Signal Corps ground augers, seven auger trucks used an ES-4 Marmon-Harrington base. The later K-44 version originated from these vehicles, even if there was no protective bodywork on the back. (U.S. Army SC-115053)

The HO-17 shelter can be qualified as a box, a box which can be displaced regardless of any motorized limits; easy to move and to set up as a fixed post, it was quite autonomous.

In 1942, thought was given to making vehicles which had to transport the Army's bulky signals equipment more rational.

The SCR-197 and its K-18 trucks and K-19 trailers were replaced by the SCR-299 which needed only the K-51 trucks to house all the equipment. But it was still too complex a concept since this radio set-up still needed a specific vehicle just to transport it. Seen from another aspect, an entire truck would become inactive once a fixed radio post was established, when it might be needed for other tasks.

The first objective was to create a mobile cabin to be installed in the cargo body of a 2½-ton, 6×6 truck like the GMC CCKW-353.

In turn, two different models of radio installation could be housed inside it, the SCR-299 which was intended for the K-51 Chevrolet, and the SCR-597 which was an improved version of the SCR-299 and specific to the U.S. Army Air Forces. The experimental HO-17-T3E1 shelter came out after various tests.

Top: Desert Training Center, 1943. The HO-17 shelter is installed here on its favorite mount, the GMC CCKW with a K-52 trailer. Uniquely, single tires have been fitted, as they are sand tires. (U.S. Army Signal Corps)

Right: Loading an HO-17-T3E1 shelter into the cargo body of a GMC CCKW-353 truck. The operation was carried out using a makeshift hoist installed on the back of a Chevrolet COE truck and a manual winch. (U.S. Army SC-172150/1)

The 2nd Armored Signal Battalion at the Desert Training Center, California, November 21, 1942. The way the shelter was designed made using it easy and adaptable. (U.S. Army SC-168178)

Above: Other views of the same vehicle make the details of this early-production shelter clearly visible, identifiable by its smooth sidewalls and simplified method of fastening it into the open GMC CCKW-353 cargo body. (U.S. Army SC-168179/168180)

When production began, it lost its suffix and became the HO-17 shelter. Made up of a base, a roof, and four panels, it was delivered as a kit and put together by the Signal Corps workshops. It slipped easily into the cargo deck of the GMC.

Its hybrid composition of wood, metal, and insulation was perfectly suited to the economies that had to be made with metal and, above all, allowed manufacture to be entrusted to companies which had not as yet put their production on a war footing.

The way subcontractors were distributed across the United States meant that a number of smaller variants

appeared in the various series of shelters, but generally they were all equipped with a rear two-part door so that the tailgate did not have to be lowered, a side door over the edge of the truck body, a simple ventilation grille on the front, and power-driven ventilation on the rear side. Its small size enabled it to be placed on the truck body which for camouflage purposes could retain its tarpaulins and bars.

Created to house the SCR-299 and its complete equipment like the K-51 truck, the HO-17 shelter finally received a more adaptable version of the SCR-299, the SCR-399, which also led to the creation of the

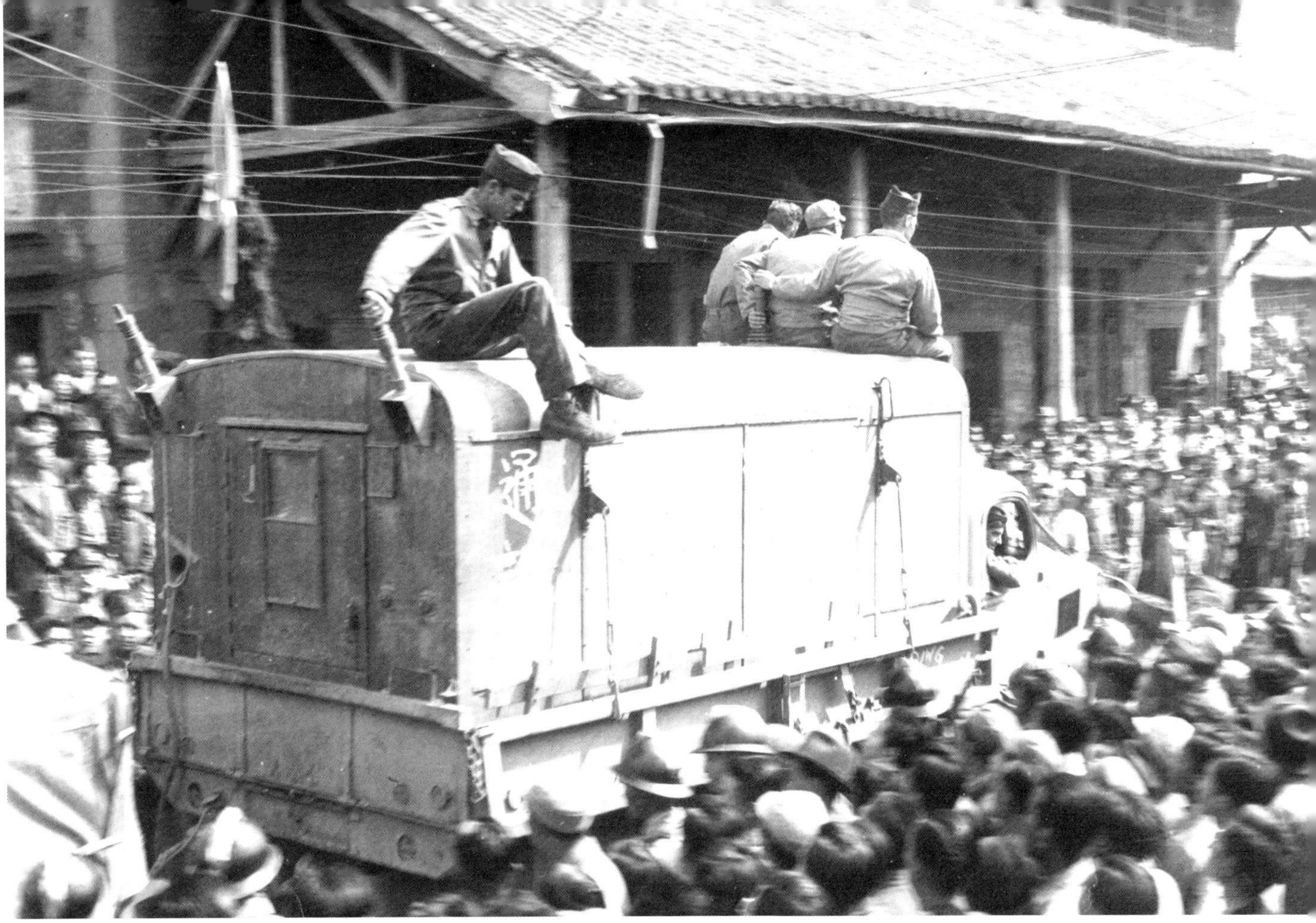

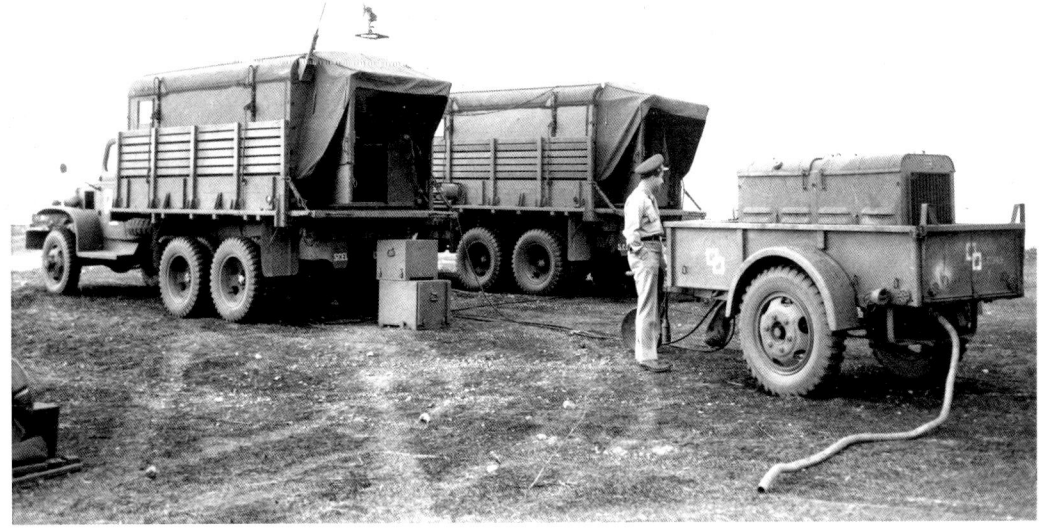

SCR-499, which was the airborne version without any shelter or trailer.

But no matter what the equipment installed in the HO-17 was, and no matter which vehicle transported it, the shelter was always accompanied by a Type PE-95 generator installed in the K-52 1-ton trailer. It was when the campaigns really got going that the versatility of the shelter became clear: it could be put on a 2½-ton cargo truck or on any other truck with a cargo floor big enough, like the Diamond-T range, or inside a GMC DUKW amphibious truck hold.

The appearance of the HO-17 shelters signaled the end of the specialist radio trucks and a new era began.

Above: January 1945, the first big overland convoy reaches China after leaving India. Greeted by a jubilant crowd, the GIs and their equipment impressed the Chinese. Among the vehicles is a GMC bearing an HO-17 shelter. (Private collection)

Left: The evolution of the materiel was constant. Photographed in November 1945, these two examples, which formed a mobile teletype set pair, made use of a protective awning for protection against the weather and allowed the caisson door to be left open for the operators' comfort. (World Wide Photo Ref. A-58055)

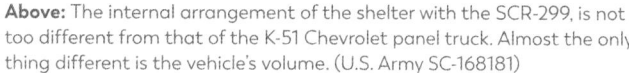

Above: The internal arrangement of the shelter with the SCR-299, is not too different from that of the K-51 Chevrolet panel truck. Almost the only thing different is the vehicle's volume. (U.S. Army SC-168181)

Above right: The standardization of the HO-17 enabled the SCR-299 to develop into the SCR-399 which was more adaptable and easier to install in the shelter. (U.S. Army SC-168182)

Right: Internal view of the left-hand-side door, its fastening system and the system for holding it open for ventilation purposes. (U.S. Army SC-168183)

Below: Many shelters were arranged for purposes other than for the SCR standards. In late 1944 on Saïpan, this one has been transformed into a message center which sorted out routine communications, both by radio and by telephone. The internal furniture is homemade. (U.S. Army SC-168185)

Opposite: Not a common use of the HO-17 shelter which is here equipped with the improved version of the SCR-299. The various components were put into chests to facilitate handling. In this form it adopted the designation SCR-399, with its registration number USA-437290. The vehicle transporting it is a GMC CCKW-353 cargo truck, part of a series of 2,946 identical vehicles, ordered in 1941 and delivered in 1942. (U.S. Army SC-C-3582)

Above: An unidentified island in the Pacific, March 25, 1945. The shelter has its main access turned to the back of the DUKW hold. The PE-95 generator has been withdrawn from its K-52 trailer to be installed transversally just behind the seats in the driver's cabin. (U.S. Army Signal Corps)

Above: Sapporo, Japan, October 13, 1945. Japanese teenagers were used as messengers for distributing messages received at or sent from this radio center. Even though the shelter was a minor technical revolution in itself, it was nonetheless not always easy to use it. Opening the rear door was only possible by lowering the vehicle tailgate and positioning steps to be able to access it. (U.S. Army Signal Corps)

Right: May 1944, in the hold of one of the numerous LSTs participating in Operation *Overlord*. In preparation for the Normandy landings, exercises and maneuvers followed each other nonstop. It was the DUKW with a radio shelter, among others, that had the important task of undertaking the radio transmissions between ships and the shore. (U.S. Army SC-293781)

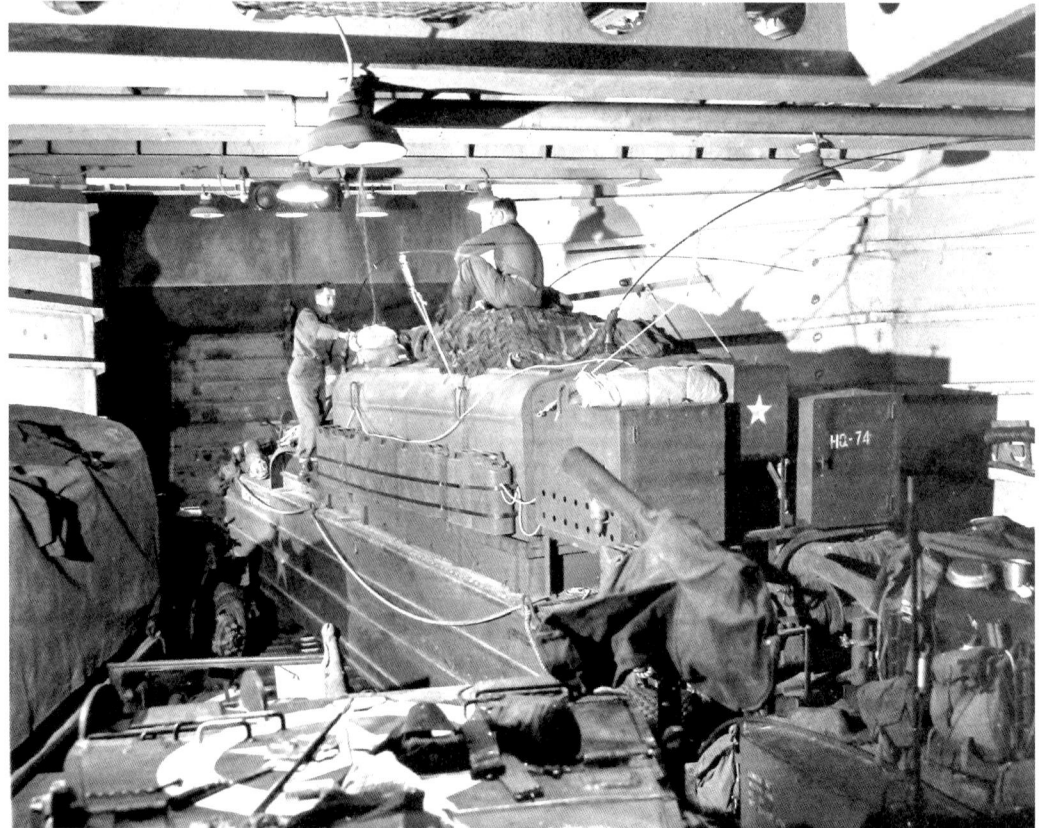

Above and below: Bristol, England, March 4, 1944. The HO-17 shelter installed in the hold of this DUKW amphibious truck. It is set up here with its front directed toward the driving seat. On the rear deck are two protective caissons for the small generators to power the SCR-299.

This signal repair truck is ready to move—doors closed, jib folded back, and winch under a cover—a design by or for the Signal Corps, but no manual mentions this vehicle. (U.S. Army Signal Corps)

Getting things in or out was not an easy matter and the doors had to be jiggled with so that nothing snagged on them. (U.S. Army Signal Corps)

The Unknown?

It is difficult, almost 80 years after the event, to provide a complete picture of the large number of developments in the area of the vehicles used for communications during World War II.

It has become almost mission impossible because some studies are still confidential. The list of the "K" models with its gaps is an excellent example of this. How can we classify the 12-volt radio Jeeps or the reel unit Jeeps when the official documentation does not mention them at all? If we consider them as prototypes, what can be said about the studies on the Chevrolet panel trucks defined as signal repair trucks or the half-track radio carrier? There were too many modifications for them to have been carried out in a simple Signal Corps workshop; the manufacturers had to be involved if they were to produce coherent and functional models. However, there was no follow-up, and no specific classification was ever given them.

Chevrolet Signal Repair Truck

Well before designing the M-30 repair trucks using the GMC CCKW-353 6×6, 2½-ton truck with the ST-6 technical body as a base, the Signal Corps created a quick response and repair vehicle for radio equipment on the base of a Chevrolet 4×4, 1½-ton panel truck.

The interior was arranged with various cupboards, shelves, and work surfaces and all the tooling needed for the work. The large, incorporated generator was so heavy and took up so much space that it had to be removed from the vehicle to operate it and to work inside the vehicle.

Handling the generator could only be done with a jib and a manual chain winch. An entire frame was fixed to the chassis to support the mast and the bodywork cut out. The right-hand side of the body was fitted for an extension ladder and a long storage chest. If one wants to liken this vehicle to any of those included in the "K" list, one just has to look at the K-70, without heed to its official designation.

Half-Track Radio Carrier

Conceived with the idea of supplying armored units with a vehicle with similar capabilities as the M2 and M3 half-tracks, four M3 half-track personnel carriers were transformed by the Signal Corps in collaboration with the Ordnance Department by replacing the armored rear part with a wooden cargo body, identical to the one mounted on the GMC CCKW-353: the idea was of installing an HO-17 shelter housing an SCR-299. As the armor on the front part of the vehicle was not vital, it was replaced by a simple sheet-metal cabin. In this form, the new vehicle was called the T-17 Half-Track Radio Carrier. But why did they use the cargo body of a GMC CCKW short-chassis truck, which could not accommodate the whole shelter?

Two other vehicles were built, this time with the cargo part of the GMC CCKW long chassis, which was better suited to the HO-17, thus becoming the T-17E1 Half-Track Radio Carrier.

Above and right: This curious combination, with a short cargo body, leaves the shelter overhanging. The comparative tests, based on the T-17's mobility and that of the GMC CCKW carrying the same load, failed to establish that the former was superior, and the project was therefore abandoned.

Sources

Primary Technical Manuals

TM 9-2700	Principles of Automotive Vehicles, 11/18/1947
TM 9-2800	Standard Military Motor Vehicles, 09/01/1943
TM 9-2800	Military Vehicles, 10/1947
TM 9-2800-1	Military Vehicles, 09/1953
TM 11-223	Signal Equipment Directory, Power Units, 04/1945
TM 11-227	Radio Communication Equipment, 04/10/1944
TM 11-241	Radio Sets SCR-197, 05/30/1942
TM 11-281	Radio Sets SCR-399-A & SCR-499-A, 03/1945
TM 11-360	Reel Units RL-26 and variants A, B, or C, 10/1944
TM 11-362	Reel Units RL-31 and variants A, B, or C, 07/01/1941
TM 11-487	Electrical Communication Systems Equipment, 02/10/1944
TM 11-1306	Technical Operation for Radio Set SCR-268, 08/1944
TM 11-1309	Technical Operation for Radio Set SCR-547, 07/1944
TM 11-1319	Operating Instruction for Radio Set SCR-527, 06/1944
TM 11-1327-P1	Technical Operation for Radio Set SCR-545, 07/25/1944
TM 11-1324	Radar Sets SCR-584, 12/1943
TM 11-1354	Radar Sets SCR-784, 03/1945
TM 11-1510	Service Manual for Radio Set SCR-270, 08/25/1944
SNL G-227	Body, Shop Truck ST and ST6, 01/1950
SNL G-508	U.S. Signal Corps Van Body, 03/1943
SNL G-685	Trailer, Telephone Cable Splicer ¼-ton Model K-38, 01/15/1945

Various technical manuals from GMC, Chevrolet, Dodge, Autocar, White, and other rolling stock

Other Publications

Crismon, Fred W., *U.S. Military Wheeled Vehicles*, Crestline Publishing

Doyle, David, *Standard Catalog of U.S. Military Vehicles*, Krause Publications

Engineering of Transport Vehicles, 1942–45 (Chief of Ordnance—Detroit) 1945

History of United States Half-Track Vehicles (Nara)

Lend-Lease Shipments World War II (Office Chief of Finance, War Department), 12/31/1946

Ordnance Department, Administrative and Tactical Vehicles, 1940–1944 (Automotive Center), 01/01/1944

Ordnance Department, Administrative and Tactical Vehicles, 1940–1945 (Automotive Center), 10/01/1945

Ordnance Department, Administrative and Tactical Armored Vehicles, 1940–1945 (Automotive Center), 01/05/1945

Radio News (U.S. Army Signal Corps), 11/1942

Radio News (U.S. Army Signal Corps), 02/1944

Summary Report of Acceptances, Tank—Automotive Material, 1940–1945 (Chief of Ordnance—Detroit), 1945

Vanderveen, B., *Historic Military Vehicles Directory,* After the Battle Publications

Vanderveen, B., *The Observers' Fighting Vehicles Directory*, Frederick Warne & Co. Ltd.

Other Titles in the Casemate Illustrated Special Series:

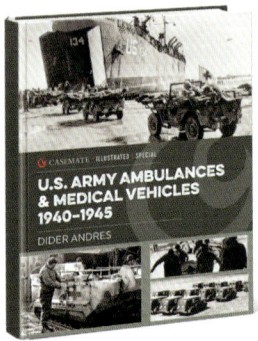

U.S. ARMY AMBULANCES AND MEDICAL VEHICLES IN WORLD WAR II

by Didier Andres

A fully illustrated book covering all types of ambulances and medical vehicles used by the U.S. Army during World War II.

JULY 2020 | 9781612008653

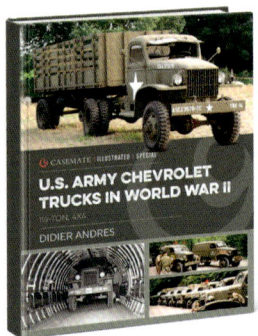

U.S. ARMY CHEVROLET TRUCKS IN WORLD WAR II: 1 ½ TON, 4×4

by Didier Andres

A fully illustrated and detailed account of the 1 ½-ton Chevy truck and its use by the U.S. Army during World War II.

MAY 2020 | 9781612008639

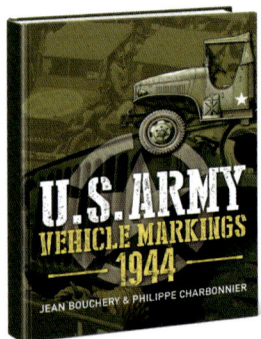

U.S. ARMY VEHICLE MARKINGS 1944

by Jean Bouchery and Philippe Charbonnier

The ultimate guide to the markings used on U.S. Army tanks, lorries and Jeeps in 1944, invaluable to historians, modelers and collectors.

JUNE 2019 | 9781612007373

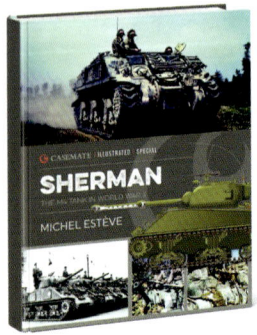

SHERMAN: THE M4 TANK IN WORLD WAR II

by Michel Estève

Fully illustrated, exceptionally detailed account of the development and deployment of the M4 Sherman in World War II.

JULY 2020 | 9781612007397